*Quantitative Methods
in the Study
of Animal Behavior*

Academic Press Rapid Manuscript Reproduction

Quantitative Methods in the Study of Animal Behavior

Edited by

BRIAN A. HAZLETT

Division of Biological Sciences
University of Michigan
Ann Arbor, Michigan

ACADEMIC PRESS   New York   San Francisco   London   1977

A Subsidiary of Harcourt Brace Jovanovich, Publishers

ACADEMIC PRESS, INC.
111 Fifth Avenue, New York, New York 10003

United Kingdom Edition published by
ACADEMIC PRESS, INC. (LONDON) LTD.
24/28 Oval Road, London NW1

Library of Congress Cataloging in Publication Data

Main entry under title:

Quantitative methods in the study of animal behavior.

    Consists chiefly of papers presented from a symposium
held at the University of Illinois at Chicago Circle,
Nov. 13, 1976 as part of the midwest regional meeting
of the Animal Behavior Society.
    Bibliography: p.
    Includes index.
    1.   Animals, Habits and behavior of—Mathematics—
Congresses.   I. Hazlett, Brian A.   II. Animal Behavior
Society.
QL751.65.M32Q36   591.5'01'84   77-21767
ISBN 0-12-335250-9

PRINTED IN THE UNITED STATES OF AMERICA

Contents

List of Contributors

Numbers in parentheses refer to the pages on which authors' contributions begin.

Wayne P. Aspey (75), The Marine Biomedical Institute, University of Texas, Galveston, Texas 77550

Catherine E. Bach (121), Division of Biological Sciences, University of Michigan, Ann Arbor, Michigan 48109

Marc Bekoff (1), Department of Environmental, Population, and Organismal Biology, University of Colorado, Boulder, Colorado 80302

James E. Blankenship (75), The Marine Biomedical Institute, University of Texas, Galveston, Texas 77550

Brian A. Hazlett (121), Division of Biological Sciences, University of Michigan, Ann Arbor, Michigan 48109

Neal Oden (203), Division of Biological Sciences, University of Michigan, Ann Arbor, Michigan 48109

June B. Steinberg (47), Department of Biological Science, University of Illinois at Chicago Circle, Chicago, Illinois 60680

Ronald S. Westman (145), Division of Biological Sciences, University of Michigan, Ann Arbor, Michigan 48109

Preface

Most of the chapters in this volume resulted from a symposium, "Quantitative Methods in Behavior," held November 13, 1976, as part of the midwest regional meeting of the Animal Behavior Society. (The methods discussed by Neal Oden in his chapter had not been fully developed at the time of the conference, but are an important addition to this volume and so his paper was included.) The meetings were held at the University of Illinois Chicago Circle campus in the Chicago Circle Center. Doctors Gary Hyatt and Robert B. Willey of the Chicago Circle campus were co-hosts for the meetings and they, along with the many graduate students and other faculty of U.I.C.C., made the meetings a profitable and pleasurable experience for all of us. Our thanks also to the University of Illinois Foundation for the funds that supported the meetings and provided honoraria for the speakers at the symposia.

When Gary Hyatt originally contacted me about chairing a symposium on quantitative methods in the study of animal behavior I was ambivalent. There was no question in my mind that such a gathering would be a good idea and one that was needed, since mathematical methods are used and misused with increasing frequency in the study of behavior. As an extreme empiricist, I have been a reasonably successful parasite of mathematicians but one might expect the chairman of a meeting to be closer to being an expert in a field than just knowing some experts. However, hopefully my level of mathematical expertise is not far below average for ethologists and this may facilitate communication with the intended audience.

While each of the papers in this collection illustrates and comments on the uses of certain mathematical approaches in studying behavior, it should be pointed out that there are several general ways in which mathematics can be an aid in understanding behavior. The most frequent uses of mathematics are attempts to deal with the variability of behaviors by statistical approaches, that is, comparisons of distributions in order to make statements on the probability that the distributions are different from one another or from some hypothesized distribution. The use of mathematics in dealing with behavioral variability allows us to consider all the data we gather. The strategy used in early ethological studies was to deduce the "main patterns" and unconsciously dismiss the variability. This of course was necessary both because of a lack of technical or mathematical tools and because the formulation of ethological principles was based upon the deduction of the "main patterns." With more statistical approaches and the availability of computers, ethologists no longer need to throw out any data—we can look for "main patterns" and the variability around these patterns. To put it another way, with appropriate mathematical techniques we can deal with the complexities of behavior, not reduce them to just the noise around simple constructs.

Of course since ethologists have a human *umwelt* (augmented by technical aids in some cases), we are always in danger of missing lots of behavioral data at the sensory level. Just as important, any good ethologist develops "central filters" as he or she learns to recognize the behavior patterns of the species under observation. Short of the cinemagraphic/analytic approach of Golani (*Behavior 44:* 89–112) we shall continue to dismiss some variability due to limitations of human observers.

The other main use of mathematics in any science is in the formulation of models or theoretical constructs. That is, if we decide a few general features of the mathematics describing behavior, then it is possible for two types of results to follow. First, there may be predictions from the model that were unexpected, that is, new ideas may emerge, which can then be tested experimentally. Second, the delineation of a mathematical model may more clearly allow falsification of the model by quantitative tests of the predictions of the model. For example, the classic Lorenz–Tinbergen ethological models have persisted as long as they have for three reasons: (a) they were intrinsically good generalizations about behavior, (b) few alternative models were proposed, and (c) clearly falsifiable hypotheses were difficult to generate due to the nature of the models. Quantitatively based models will, in general, have a shorter half-life than other models since they can be falsified more quickly.

The papers in this volume discuss these two uses of mathematics. The chapters by Wayne Aspey, June Steinberg, Neal Oden, and Marc Bekoff are primarily concerned with statistical approaches—techniques by which we can describe the main patterns in *and* the variability of the behavior patterns we study. The chapters by Ron Westman and Brian Hazlett deal primarily with model building. It is hoped that this mixture of approaches will provide something of interest to a variety of readers.

I wish to express my thanks to Catherine Bach, my wife, for her aid and tolerance during the editing of this book, to the staff of Academic Press for their cooperation and encouragement, and to Pat Wander for the careful typing of the manuscript.

QUANTITATIVE STUDIES OF THREE AREAS OF CLASSICAL ETHOLOGY: SOCIAL DOMINANCE, BEHAVIORAL TAXONOMY, AND BEHAVIORAL VARIABILITY

Marc Bekoff

University of Colorado

Abstract: Among the areas of classical ethology that still constitute a large portion of current research in behavioral biology are social dominance, the use of behavioral characters in the assessment of taxonomic relationships, and behavioral variability and stereotypy ("fixed" action patterns). Unfortunately, until recently, there have been very few quantitative analyses of these classical areas. In this chapter, I shall consider some ways in which these three topics have been analyzed quantitatively. Specifically, I shall discuss (1) the application of Landau's index to the measurement of social dominance hierarchies, (2) the use of procedures that are commonly employed in numerical taxonomy to assess taxonomic relationships based on non-behavioral characters, and how some of these have recently been applied to behavioral taxonomies, and (3) the way in which behavioral variability and/or stereotypy have been measured, primarily by use of the coefficient of variation.

Modern ethology has advanced to a level at which quantitative analyses must be forthcoming in these and other areas of research. Yet, the "realities" of any given behavioral system, organism, and research problem must not be obscured by using techniques for which the underlying assumptions cannot be met. That is, there must be congruence between the problem at hand and the analytical method that is employed.

...biology is a system that proceeds from biochemistry to the associated subjects of neurophysiology and genetics. All else, as they used to say of the non-physical sciences, is stamp-collecting (de Solla Price, 1960).

All too often ethologists paint a vivid picture of the

behavioral element of an animal without giving a single
figure as to its probability of occurrence or duration;
psychologists present meticulous quantified results
subjected to analyses of variance while giving only the
scantiest indications of how the animal behaves (Hutt
and Hutt, 1970).
In my own surroundings, I notice that those who are
most positive in the matter of (these) difficult
questions are those who have seen the least (Fabre,
1916).
In few fields of biological study is it so easy to obtain results and so difficult to explain them as it is
in the study of animal behavior (Simpson, 1969).
But we must be wise and careful in applying the
theorems and results of communication theory, which are
exact for a mathematical ergodic source, to actual
communication problems (Pierce, 1961).

INTRODUCTION

In this paper, I shall consider the following three topics: social dominance, behavioral taxonomy, and behavioral
variability (or stereotypy). I do not plan to review each of
these fields other than to give a brief background so that
the quantitative techniques that are discussed will be understood with respect to the behavior under consideration. For
discussions of sampling methods the reader is referred to
Altmann (1974) and Dunbar (1976). My purposes are specifically to discuss a variety of ways in which ethologists may
analyze some "classical" concepts and to present some recent
examples that have approached the above classical concepts
using one or another quantitative techniques. I have provided
a rather lengthy reference section to allow the reader to see
how various analytical procedures have been applied to these
behavioral problems among a wide variety of animal species.
In the behavioral sciences, it usually is the case that
qualitative descriptions of behavioral phenotypes give rise
to quantitative analyses. Certainly, this is a healthy trend.
However, quantitative "overkill" may be damaging, and the
temptation to make the leap from the initial compilation of
descriptive information to the use of highly sophisticated
mathematical analyses requires careful consideration. We
should choose an analytical procedure with the same degree of
care that we exercise when we choose a piece of equipment to
increase our powers of observation. With regard to quantitative procedures, we must be very careful since there are
often many underlying assumptions that must be met and in
many cases the ability of animals to behave in a manner that

would allow the assumptions to be supported is at best ten-
uous. This is particularly true when sequential variability
is the problem under study (see below). As Slater (1973) has
stressed, it is a simple matter -- the less valid the assumptions, the more unrealistic are the results. The degree to
which researchers choose to overlook certain guidelines for
the use of various methods, and the effect of these "over-
sights" on the validity of their results and conclusions, is
difficult to assess (Siegel, 1956). Although "slight" deviations on meeting assumptions may not have radical effects,
there seems to be no general agreement as to what constitutes
a "slight" deviation (Siegel, 1956, p. 20).

In light of the current interest in quantitative
ethology, it is refreshing to see that some researchers (many
of whom have equal expertise in the behavioral sciences as
well as in mathematics) are professing the use of extreme
restraint in the application of various analytical procedures
(e.g., discussion of Cane, 1959; Chatfield and Lemon, 1970;
Simpson, 1973; Slater, 1973). Just because something has been
done before does not mean that it is correct, and the wide
availability of "canned" computer programs should not be the
factor determining how a problem(s) should be analyzed or in
ultimately directing further research. The suggestion that
seems to be the most appropriate is, simply, fit the method
of analysis to the animal, not the animal to the method. Do
not make the animal something that it is not (e.g., a
stationary beast; see below); do not "short-circuit"
evolution! Lest I lead you to believe that I am going to
conclude that quantification of ethological data is impossible
and that behavioral research and stamp-collecting are
synonomous (an insult to serious philatalists), let me now
consider how the problems of social dominance, behavioral
taxonomy, and behavioral variability have been analyzed or
can be studied using standard quantitative techniques for
which the underlying assumptions can be satisfied in many
ethological endeavors. I have chosen these three topics
because there is a good deal of current research being conducted within these areas, they are truly "classical" concepts, and because of my familiarity with them.

I. SOCIAL DOMINANCE

The concept of social dominance, or status hierarchies,
has received a lot of attention ever since Schjelderup-Ebbe
(1922, translated in Schein, 1975) published his work on the
social psychology of domestic chickens in which he described
the social organization, or pecking orders, of these birds.
Within the last decade or so, detailed analyses of "dominance"

and, in particular, the sweeping generalizations that accompany the idea, have been called into question by a number of investigators (e.g., Rowell, 1966, 1974; Brantas, 1968; Gartlan, 1968; Bernstein, 1970; van Kreveld, 1970; Watson and Moss, 1970; Drew, 1973; Kummer, 1973; Spigel and Fraser, 1974; Clutton-Brock, Greenwood, and Powell, 1976; Hanby, 1976; Marler, 1976). The word itself is used in a wide variety of contexts (Gartlan, 1964, 1968; Rowell, 1974), with definitions including information about competitive abilities (e.g., priority of access) and traits that are related to an individual's status, and even going as far as using the terms "aggressive" and "dominant" as synonyms. Aggressive individuals are not necessarily the most dominant in a group, aggression does not necessarily lead to the establishment of dominance relations, and the formation of a dominance hierarchy does not necessarily lead to a reduction in aggression (e.g., Wolfe, 1966; Gartlan, 1968; Rowell, 1974; Marler, 1976; Potter, Wrensch, and Johnston, 1976). Furthermore, a number of studies have indicated that there is often no correlation between different behaviors that are thought to be correlated with dominance. That is, social rankings based on different criteria frequently do not correlate well with one another (Gartlan, 1968; van Kreveld, 1970; Bernstein, 1970; Drew, 1973; Rowell, 1974; Syme, 1974; Lockwood, 1976; but also see Richards, 1974 and Clutton-Brock and Harvey, 1976 for alternative views). Suffice it to say, the concept of dominance still is in need of refinement.

Although global definitions of dominance are difficult to come by, it is a fact that in a wide variety of animal societies, individuals comprising a group can often be ranked on a dominance-subordinate scale and specific behavior patterns and physiological states can be associated with relative ranks among the individuals (e.g., priority of access to estrous females (see DeFries and McClearn, 1970; LeBoeuf, 1974; Hausfater, 1975; and Hanby, 1976 for discussions); the control of spatial relationships (McBride, 1964; Watson and Moss, 1970; Myers and Krebs, 1971, 1974; Bekoff, 1977a,b); activity (Dunbar and Crook, 1975); rate of ultrasound vocalization (Nyby, Dizinno, and Whitney, 1976) and electric organ discharge (Westby and Box, 1970; Bell, Myers, and Russell, 1974); altered maturation (Sohn, 1977); changes in heart rate and other physiological parameters (Candland, et al., 1969, 1970, 1973; Cherkovich and Tatoyan, 1973; Manague, Lesher, and Candland, 1975)). In this section, I shall briefly discuss one particular way in which dominance hierarchies can be "measured" with respect to their degree of linearity. Landau's (1951a,b, 1953, 1965; reviewed by Chase, 1974) method will be considered since it is applicable to studies of social dominance in a very general way. Since the concept of dominance and all of its ramifications are still very much a part of modern day behavioral science, it is essential that more accurate descriptions be forthcoming.

A. Types of Hierarchies

There are three major types of social hierarchies. The first is a despotism in which one individual dominates all other members of his or her social group, with no rank distinctions among the subordinates (Wilson, 1975). This has been observed in iguanid lizards living under conditions of unnaturally high densities (Carpenter, 1971, cited in Wilson, 1975). Most usually, hierarchies are referred to as being linear or non-linear. In the first case, an individual (usually called the alpha animal) dominates all other group members, another individual (beta) dominates all group members but alpha, and so on. In order for a linear hierarchy to exist, two conditions have to be fulfilled: (1) the dominance relations must be asymmetric, that is, for every paired interaction, one individual can be classified as being dominant and (2) dominance relations must be transitive (Figure 1 A,C; see Harary, Norman, and Cartwright, 1965, Chapter 11; and Moon, 1968, pp. 14ff), that is for any three animals if 1 dominates 2 and 2 dominates 3, then 1 also dominates 3. Non-linear hierarchies, on the other hand, are those in which there is at least one or more intransitive triads (Figure 1 B,D; Chase, 1974). Chase (person.

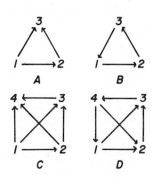

Fig. 1. Examples of linear (A,C) and non-linear (B,D) social hierarchies. The arrow means "dominant over". Note that in A and C, there are no intransitive triads. However, in D there is an intransitive triad among individuals 1, 3, and 4. That is, 1 dominates 3, 3 dominates 4, but 4 dominates 1. There is also an intransitive triad among individuals 2, 3, and 4 (adapted from van Kreveld, 1970).

comm.) further points out that there must be com-
pleteness (i.e., there must be a relationship between all
pairs of animals).

In both linear and non-linear hierarchies, it is possible
to have different types of "pecking" relationships
(assertions of dominance) between pairs of individuals. If
assertions of dominance go only in one direction, the term
"peck-right" is frequently used as a shorthand description.
This type of relationship can exist in both linear (pecking
only goes "down" the hierarchy) and non-linear hierarchies
(e.g., an intransitive triad). If, however, the situation
is encountered in which two animals "peck" at one another in
unequal amounts so that one can nonetheless be classified as
being dominant, a "peck-dominance" relationship exists.
Peck-dominance hierarchies can also be linear or non-linear
(see Table 1).

B. The Application of Landau's Method

Landau developed a method by which the degree of
linearity of a social hierarchy could be measured and stan-
dardized for intergroup comparisons. An index of linearity,
h, is calculated using the following formula

$$h = \frac{12}{n^3 - n} \sum_{a=1}^{n} \{v_a - (n-1)/2\}^2$$

where n = the number of animals in the group and v_a = the
number of individuals that an individual dominates. The term
$12/(n^3-n)$ normalizes h so that values range from 0 to 1.
When h = 1, the hierarchy under study is linear and the
variance of v_a is at its maximum. When h = 0, each animal
dominates an equal number of group members and v_a = 0. An h
value of 0.9 or greater may be taken to mean that the hierarchy is "strong", that is, it is approaching linearity (Chase,
1974). Although Landau's method does not generate a statistic
from which a level of significance (p-value) can be calculated
to determine whether or not an observed hierarchy differs
from that which would be expected by chance, other workers
have developed a statistic (that generates a chi-square value)
that is very similar to Landau's in form (David, 1959).
However, the statistical measure of linearity only applies
when group sizes are greater than 6. For the moment, at
least, using h ≥ 0.9 as a "cut-off" for separating strong,
nearly linear hierarchies, from non-linear hierarchies
seems appropriate (see Chase, 1974).

Although Landau's method for assessing the degree of

linearity of a social hierarchy(ies) has been in the literature for approximately 25 years, it has been used only rarely
(Chase, 1974; Lockwood, 1976). Chase (1974) has applied the
technique to data collected on the formation and stability of
pecking orders in birds, however, there do not appear to be
any published reports for mammals. I would like to consider
two examples in which the application of Landau's formula has
been useful. The first set of data comes from ontogenetic
studies on infant coyotes (Canis latrans) and Eastern coyotes
(C. l. var; see section on Behavioral Taxonomy for the
rationale behind separating these animals from one another).
The second example stems from the work of Lockwood (1976) on
pack structure in captive wolves (C. lupus).

C. Hierarchy Formation in Coyotes

     Coyotes and Eastern coyotes engage in rank-related
agonistic encounters during the 3rd to 5th weeks of life
(Bekoff, 1974, 1977a, In Press$_a$, Bekoff, Hill, and Mitton,
1975). Intense fights are extremely common and clear-cut
dominant-subordinate relationships are established as a result
of these early interactions. Data relevant to hierarchy
formation are presented in Table 1. Full litters of animals
were observed and patterns of interaction among all litter-
mates were studied. In all three litters, a perfect linear
hierarchy was established (h = 1). Note that there is also a
direct relationship between the size of the litters and the
length of time that it took to attain linearity. For example,
in coyote litter 2 (n = 4), after only 3 days of fighting a
linear dominance hierarchy was established. It should also be
stressed that a linear hierarchy may remain linear in form
even if some animals shift in rank. In coyote litter 1, for
example, linearity was maintained from the second to the ninth
month of life although there were some shifts in relative
rank among the individuals during the first 90 days. From
days 50-90, h = 1, with the exception of those days on which
a reversal of dominance occurred, and "acceptance" of
subordination required a series of agonistic encounters. For
further details concerning the development of social relationships in canids, see Bekoff (1974, 1977a, In Press$_b$). In
particular, the necessity of my determining social rank for
individual coyotes comprising a litter and overall hierarchical structure is amplified in Bekoff (In Press$_a$).

D. Hierarchies in Wolf Packs

     The only other mammalian species for which I could find

TABLE 1

Hierarchy formation in infant canids analyzed using Landau's index (h). All of the hierarchies are examples of peck dominance hierarchies. One animal was assumed to be dominant over another individual if it "won" over 70% of the agonistic encounters in which there was a clear outcome. Among these young canids, it is not difficult to determine "winners" and "losers".

	Days of Age	Number of Fights	h
Coyote litter 1* (n=6)	25-31	220	.77
	32-35	59	.88
	36-50	50	1.00
Coyote litter 2 (n=4)	23-25	40	1.00
	26-29	38	1.00
	30-35	36	1.00
Eastern Coyotes (n=5)	21-25	65	0
	26-29	178	.80
	30-35	137	1.00

* Coyote litter 1 was mother-reared and the other two litters were hand-reared

data concerning a measurement of linearity for observed hierarchies is the wolf, C. lupus (Lockwood, 1976). The results of this study are presented in Table 2. The interesting aspect of this analysis, and the reason for its inclusion, is that Lockwood analyzed not only dominance hierarchies but also subordinate hierarchies as well. By looking at both sides of the coin in dominant-subordinate relationships, the importance of the subordinate individual to the formation and maintenance of social hierarchies is stressed (see further discussions of this idea by Rowell, 1966, 1974; Lockwood, 1976). Note that Lockwood found no "strong" dominance hierarchies and also that the subordinate hierarchy is stronger in one-half of the cases, and even is linear in one instance.

The above examples demonstrate that Landau's index of linearity is applicable and useful in the analysis of social relationships in groups of animals. Calculation of "h" is straightforward and the measure can be used in studies in which different groups are compared to one another. That is, a value of h is a standard that means the same thing for different groups. It simply is an indication of the degree of

TABLE 2

The Linearity of Dominance and Subordinate
Hierarchies in Captive Wolves (from Lockwood, 1976)

Wolf Pack	$h_{dominant}$	$h_{subordinate}$
350-72 (n=5)	.10	.10
350-73 (n=4)	.30	1.00
350-74 (n=4)	.30	.30
OM-73 (n=4)	.50	.30
Big Pack (n=7)	.69	.43
Satan (n=5)	.20	.25
Release (n=5)	.20	.80
Mix-72 (n=6)	.65	.09
Mix-74 (n=6)	.17	.31
Jinx (n=6)	.54	.57

linearity of a social hierarchy. Lastly, Landau's formula can be used to analyze dominance relations that result from inter- actions between and among individuals comprising whole animal groups. This is important, since dominance hierarchies that are based on the results of experimentally controlled, paired, round-robin interactions frequently do not represent what actually is happening in the real world of social groups of animals (Chase, 1974).

II. BEHAVIORAL TAXONOMY

Since the turn of the century, it has been recognized that behavioral characters can be used to assess and/or estab- lish taxonomic relationships. This idea was laid out clearly in the work of Whitman and Heinroth and extended by Lorenz and his students. Implied in this belief are the notions that behavior, like any other character, evolves, and accordingly, that the "structure" of behavior has a traceable phylogeny. Precisely tracing the phylogenetic history of a behavioral phenotype is difficult or impossible since behavior does not fossilize. However, the use of extant organisms has provided some interesting and important data concerning the use of behavior in the assessment of taxonomic affinities (for review see Lorenz, 1941; Marler, 1957; Mayr, 1958; Cullen, 1959;

Atz, 1970; Eibl-Eibesfeldt, 1975).

There have been a considerable number of studies in which a wide variety of behavioral characters have been used to analyze taxonomic relationships (Table 3). Behavioral taxonomies have given support to taxonomic schemes based on non-behavioral characters or have been useful in clarifying discrepancies. Nonetheless, among the wide variety of studies that have been conducted, extremely few have had a strong quantitative base. I would like to consider a few examples that exemplify different approaches to the problem(s).

A. "Yes"-"No" Approaches

Earliest studies of behavioral taxonomy used the yes-no approach. If an animal performed a certain behavior pattern, a "yes" was recorded and conversely, if the same pattern was not observed for a given species, then it received a "no". The first major attempt to apply this checklist method to a group of closely related animals was performed by Lorenz (1941; translated in 1971). Lorenz studied 17 species and three genera of ducks and geese, Family Anatidae. The results of his work are summarized in his now classic diagram (Figure 2) in which the different species are linked together by shared characters. Lorenz's "anatidogram" (G. Barlow, pers. comm.) is useful in that evolutionary relationships are clearly pictured and the relationship between behavioral as well as non-behavioral characters is presented.

Another application of the yes-no method is presented by Dewsbury (1972). Dewsbury studied the copulatory patterns of male mammals and asked a series of yes-no questions for four major aspects of male reproductive behavior (Figure 3). Different species could then be classified into one of 16 patterns. Using this approach, Dewsbury (see his Table 1) found that closely related animals tended to fall within the same pattern, and taxonomic affinities were detected.

Although yes-no questions are useful in studying taxonomic relationships among animal groups, more rigorous quantitative procedures are available for behavioral taxonomic endeavors, particularly when the differences between or among groups are quantitative and not simply qualitative. These techniques are reviewed in Sneath and Sokal (1973) and have, to date, been used almost exclusively to assess taxonomic relationships using non-behavioral characters. The application of many of these procedures is entirely appropriate for behavioral studies.

TABLE 3

Some Studies in Which Behavioral Characters
Were Used in the Analysis of Taxonomic Relationships

Authorities	Animal Group (and Criterion(a))
Heinroth (1910); Whitman (1919)	Drinking by pigeons
Petrunkevitch (1926)	Spiders: analyses of "instincts"
Lorenz (1941, 1958)	Anatidae (ducks and geese); general behavior patterns
Delacour and Mayr (1945)	Anatidae
Spieth (1952)	Drosophila courtship
Barber (1953)	Fireflies of the genus Photoris; flash characteristics
Schmidt (1955)	Termite nests
Strokes (1955)	Gall midges; host plant choices
Andrew (1956a)	Passerine birds; flight intention movements
Sibley (1957)	Various birds
Simmons (1957)	Head-scratching by birds
Tinbergen (1959, 1960)	Gull displays
Dilger (1960,1962)	African parrots of genus Agapornis
Johnsgard (1961)	Anatidae
Kaston (1964)	Spider webs
McKinney (1965)	Anatidae; comfort movements
van Tets (1965)	Pelican displays
Salthe (1967)	Salamandridae courtship
Littlejohn and Oldham (1968)	Rana pipiens mating calls
Lomax (1968); Lomax with Berkowitz (1972)	Human song and dance patterns
Struhsaker (1970)	Cercopithecus monkeys; vocalizations
Echelle, Echelle, and Fitch (1971)	Anolis; aggressive displays
Brown and Brown (1972)	Rana pipiens vocalizations
Cattell, Bolz, and Korth (1973)	Domestic dogs
Rovner (1973)	Wolf spiders; copulatory patterns
Berg (1974)	Strombid gastropods; feeding, locomotion, righting, predator escape
Michener (1974)	Bee sociality
Bekoff, Hill, and Mitton (1975)	Canids; patterns of social development
Dixson, Scruton, and Herbert (1975)	Talapoin monkeys (and other Old World Catarrhines); grooming invitations, facial expressions
Dunford and Davis (1975)	Chipmunk vocalizations
Heymer (1975)	Dragonflies
Vierke (1975)	Fish (Belontidae); reproductive and parental behavior

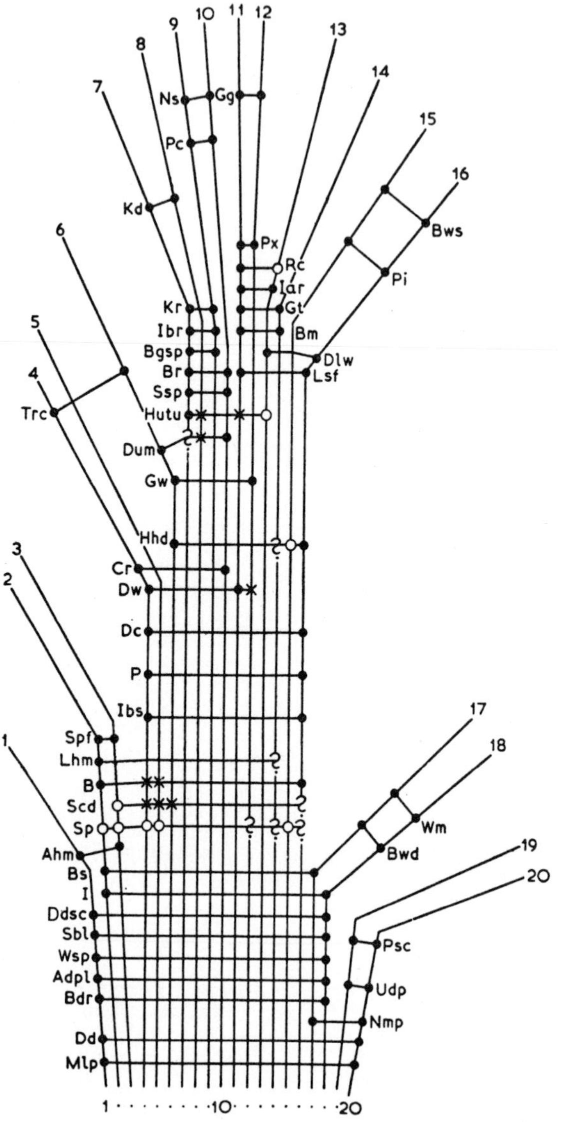

Fig. 2. A taxonomic scheme ("anatidogram", G. Barlow, personal communication) of ducks and geese based on behavioral characters (from Lorenz, 1941, 1971). The vertical lines represent species; the horizontal lines characters common among them. A cross indicates the absence of a character in a species crossed at the point concerned by a character cross-line. A circle indicates special emphasis and differentiation of the character. A question-mark in-

*dicates Lorenz's uncertainty. The few morphological charac-
ters that were used showed the similarity in their distribution when compared to the behavioral characters. For an
explanation of the characters and identification of the
species see Lorenz (1971). Reproduced with permission from
Lorenz, 1971; Methuen Ltd., London and Harvard University
Press, Cambridge, Mass.*

B. Quantitative Approaches

There are very few examples of a behavioral taxonomy
that is based on rigorous analytical procedures (Andrew,
1956a; Cattell, Bolz, and Korth, 1973; Michener, 1974; Bekoff,
Hill, and Mitton, 1975; Breed, 1976). Recent work on bees and
canids can be used to demonstrate the application of two standard procedures of numerical taxonomy to behavioral taxonomic

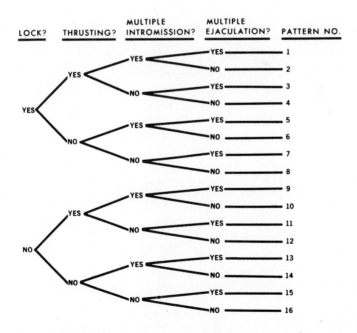

*Fig. 3. Patterns of copulatory behavior in male mammals
(reproduced with permission from Dewsbury, 1972).*

endeavors. Details on methodology can be found in the original sources and in greater detail in Sneath and Sokal (1973).

C. Principal Components Analysis of Bee Sociality

Michener (1974) used a series of 28 behavioral characters (Table 4) to analyze the taxonomic relationships among 18 species of bees. His data were analyzed using a technique known as principal components analysis (see also Dudzinski and Norris, 1970; Aspey, this volume) which is a type of factor analysis that reduces the dimensionality of a multivariate data set to fewer (usually two or three) dimensions. The first principal axis or eigenvector is that dimension accounting for the greatest amount of variance of the sample, the second principal component (axis) accounts for the second largest amount of variance and so forth. Usually, only enough axes (eigenvectors) are extracted to account for 75% of the variance of the total variance of the data set (Sneath and Sokal, 1973).

Michener (1974) graphed his data on a three-dimensional system of coordinates (Figure 4). In this way, each species can be represented as a point in space that indicates where the species falls with respect to all of the other species, using the chosen behavioral characters. As can be seen from Figure 4, the 18 species of bees were divisible into four

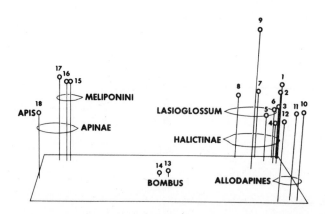

Fig. 4. A three-dimensional scatter diagram showing the positions of 18 species of bees with respect to the first three principal components extracted (see also Table 4 for some of the characters used in this analysis; original drawing by B. Siler; reproduced with permission from Michener, 1974, Harvard University Press, Cambridge, Mass.).

TABLE 4

Some of the 28 Characteristics (Variables) Related to Social Level and States that Michener (1974) Used in his Behavioral Taxonomic Analysis of 18 Species of Bees

Variables or Characteristics	States
Maximum normal number of females per nest	(1) 1; (2) 2-19; (3) 20-199; (4) 200-1999; (5) >1999.
Sex ratio in early colony growth (that is, seasonal changes in male production)	(1) Normal; (2) few males; (3) no males.
Queen ovarian specialization	(1) Ovaries as in solitary relatives (solitary forms placed here); (2) ovaries > than in solitary relatives; (3) ovaries >> than in solitary relatives; (4) ovaries > than and with many more ovarioles than in solitary relatives.
Worker fertilization	(1) Regularly fertilized; (2) < 50% fertilized; (3) rarely fertilized; (4) never fertilized.
Worker layer of reproductive eggs	(1) Unfertilized bees with queenlike ovaries among the workers; (2) of regular occurrence, but workers always less productive than queens; (3) uncommon, only occassional worker laying (4) absent in presence of queen.
Larval food	(1) Mostly honey and pollen (2) major glandular component; (3) mostly of glandular origin.
Eggs and larvae closed in cells	(1) Yes; (2) no.
Food storage for adults	(1) Absent; (2) well developed.
Food exchange among adults	(1) Absent; (2) via brood cells or brood food only; (3) via storage pots; (4) direct and uncommon; (5) direct and extensive.
Removal of waste from larval quarters before pupation	(1) No; (2) yes.
Secretion of cell construction material	(1) None; (2) inner lining; (3) major constituent mixed with pollen or resin; (4) entirely secreted.
Communication as to food sources	(1) Unknown; (2) social facilitation including odor; (3) odor trail and leading; (4) partial leading; (5) dance communication.
Defense at entrance	(1) Irregular and no obvious guarding; (2) constricted entrance and one guard; (3) larger entrance often with several guard bees.
Temperature control (warming)	(1) Absent; (2) individual or group incubation; (3) group warming or large area.
Alarm pheromones	(1) Unknown and probably absent; (2) present.
Survival through unfavorable seasons	(1) Fertilized females usually alone; (2) colony.
Response to emergency conditions such as starvation	(1) Inactivity, death; (2) absconding.

discrete clusters. Furthermore, and of greater importance, is the fact that although Michener had used behavioral characters specifically related to levels of sociality, the four clusters represented previously established taxonomic units rather than social levels. Here, is a fine example of concordance between taxonomic relationships based on behavioral as well as non-behavioral characters.

D. Two Approaches to a Behavioral Taxonomy of Canids

    It is a well established fact that even among closely related canid (family Canidae) species there are distinguishing behavioral characteristics (Scott and Fuller, 1965; Kleiman, 1967; Bekoff, 1975, In Press$_a$). This is even so for domestic dogs (C. familiaris), all of whom are classified within the same species (Scott and Fuller, 1965; Cattell, Bolz, and Korth, 1973; Fox and Bekoff, 1975). Cattell, Bolz, and Korth (1973) studied 101 dogs representing five different breeds. They used 42 behavioral and somatic characters and reduced these to 16 factors. Of the 16 factors, 15 referred to behavioral characters and the 16th contained all of the physical variables. In order to decide whether or not behavioral characters could separate the five breeds of dog, Cattell et al. used a statistic called the pattern similarity coefficient, r_p which gives a measure of the likeness of the two profiles of independent measures (Cattell, 1949).

$$r_p = (2k - \Sigma^n d^2) / (2k + \Sigma^n d^2), \text{ where } k = \chi^2_{(0.50,n)},$$

n = the number of variables in the profile or the degrees of freedom, and d is the difference between variable scores for the two individuals. Cattell et al. found a separation among the dogs that roughly corresponded to the different breeds. They point out that it would be interesting to apply their technique to groups in which the taxonomic relationships were unknown, rather than merely attempting to match an existing classification scheme.
    Although the pattern similarity coefficient measure has not been used in such a study, another method, discriminant function analysis, has been used to study the taxonomic relationship of an "unknown" canid with known coyotes and wolves. One advantage of the comparative approach to behavior is that it allows an investigator to analyze in detail the similarities and differences among a number of different species. Scott (1967) suggested that one thing that distinguishes different members of the family Canidae from each other is the differential development of social behavior, and among the members of the genus Canis that have been studied in considerable detail (Bekoff, 1974, In Press$_a$), this is indeed the

case. Since young <u>Canis</u> show distinct and significant
differences in behavioral ontogeny, we decided to use behavioral phenotypes as analyzable characters to assess the
taxonomic relationships among infant coyotes, wolves, and an
"unknown" canid (Bekoff, Hill, and Mitton, 1975) that has
been called the New England canid or "Eastern coyote".
Specifically, we analyzed the time-course of development of
agonistic behavior and social play. All of these species
have the same basic behavioral repertoire (Bekoff, 1977a),
and the one factor distinguishing them from one another is the
relative frequency of occurrence of rank-related agonistic
behavior (see above) and social play (Bekoff, 1974, In
Press$_a$). Infant coyotes are significantly more aggressive and
less playful than the same-aged, infant wolves. The major
question of interest concerned how the "unknown" canid would
compare behaviorally with coyotes and wolves, and whether or
not there would be agreement between our taxonomic scheme
based on behavior and taxonomic relationships suggested by
analyses of skeletal and dental measurements (Lawrence and
Bossert, 1969, 1975).

In order to make a direct comparison with the earlier
work of Lawrence and Bossert, we chose to use the same
technique that they employed, namely, a linear discriminant
function analysis (see also Mihok, 1976). Briefly, linear
discriminant function analysis is a method by which the means
of various characters from two populations are compared to
one another. The resulting number, or discriminant function,
may be thought of as a weight for each character when it is
pooled with all other characters. This weight indicates the
degree of separation of the two populations -- the larger the
number, the greater the separation. When the results are
plotted on a linear axis, it is also possible to fit a third
"unknown" group to determine where it falls with respect to
the other two populations (Figure 5).

The results of our analysis are presented in Table 5 and
Figure 5. During this study, we analyzed a total of 4227
interactions (see Bekoff, Hill, and Mitton, 1975 for details).
A total of four behavioral characters were analyzed (Table 5).
The differences among the groups that had been documented
in earlier studies were now weighted, and inspection of Table
5 shows that coyotes, when compared both to wolves and New
England canids were more aggressive than the wolves and less
playful. When all of the data were combined and plotted on a
linear discriminant function axis (Figure 5, top), the New
England canids fell intermediate to the wolves and coyotes but
closer to the coyotes (mean positions for the wolves, New
England canids, and coyotes were -57.88, -8.90, and +14.14,
respectively). A pair-wise analysis in which each group was
matched with one another (Figure 5, bottom) showed that the

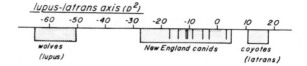

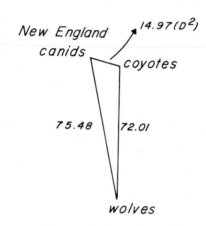

Fig. 5. The results of a behavioral taxonomy study on infant coyotes, wolves, and New England canids (=Eastern coyotes). The relative frequencies of occurrence of social play and agonistic behavior were used as behavioral characters (Bekoff, Hill, and Mitton, 1975). Top: Linear discriminant values of known C. lupus (wolf) and C. latrans (coyote) litters cast on a lupus-latrans discriminant axis onto which New England Canids are projected. Note that the New England canids fall between lupus and latrans, but closer to latrans. Bottom: Distances (D²) in discriminant function units based on pairwise analyses of lupus, latrans, and New England canids. Note the close relationship between coyotes and New England canids and that both fall approximately the same distances from wolves.

coyotes and New England canids were more closely related to one another than either was to the wolves. Lawrence and Bossert (1969, 1975) showed a similar relationship using skull and dental characters. Accordingly, the use of the term Eastern coyote (C. latrans var.) is justified both on behavioral and morphological grounds (see also Silver and Silver, 1969).

In summary of this section, it should be clear that behavioral taxonomies may be quantitatively analyzed using a

TABLE 5

Discriminant functions differentiating species of canids on the basis of four behavioral characters. The top numbers are weights for the characters and define the axis that discriminates to the greatest degree between the species. The higher the number, the better the discrimination. The numbers in parentheses indicate the observed proportion of the behavior in relation to the total number of interactions observed in the stated time period for each species pair. Ranges of proportions are reported for the ten pairs of New England canids. For example, 0.47 and 0.25 are the proportions of observed agonistic behavior (character 1) for the two pairs of coyotes, and 0 and 0 are the corresponding proportions of agonistic behavior for the two pairs of wolves (from Bekoff, Hill, and Mitton, 1975).

Combination	Agonistic behavior		Play behavior	
	Character 1 (days 21-28)	Character 2 (days 29-35)	Character 3 (days 21-28)	Character 4 (days 29-35)
Coyote/wolf	46.61 (0.47,0.25/ 0,0)	0 (0.36,0.27/ 0.11,0.09)	312.99 (0.01,0/ 0.21,0.16)	0 (0.14,0.09/ 0.34,0.30)
Coyote/New England canid	9.60 (0.47,0.25/ 0.09-0.32)	35.09 (0.36,0.27/ 0.07-0.23)	110.31 (0.01,0/ 0.03-0.10)	65.88 (0.14,0.09/ 0.11-0.20)
Wolf/New England canid	63.75 (0,0/ 0.09-0.32)	56.92 (0.11,0.09/ 0.07-0.23)	116.56 (0.21,0.16/ 0.03-0.10)	297.20 (0.34,0.30/ 0.11-0.20)

variety of rather standard techniques that have been used in taxonomic endeavors based on non-behavioral characters. Just as with measurements of social dominance hierarchies, the application of suitable quantitative techniques will provide a more solid base from which to generate further research ideas and hypotheses.

III. BEHAVIORAL VARIABILITY AND STEREOTYPY

Interest in the variability (and stereotypy) of patterns of motor behavior has long permeated ethological thought. Indeed, the identification of species-specific action patterns lead Whitman (1898-99) and independently, Heinroth (1910), to suggest that behavioral characters or phenotypes could be used to assess taxonomic affinities among animal groups. Lorenz, a student of Heinroth, extended the earlier work of both Whitman and Heinroth by stressing that behavior patterns are something that an animal "has got", similar to anatomical structures such as claws, teeth, and internal organs, and as such, could be studied from a comparative point of view (see above section). One problem, however, that has not received much attention until fairly recently, concerns the measurement of behavioral phenotypes, particularly their form or "morphology". In this section, I shall consider behav-

ioral variability from two points of view. First, I shall
discuss ideas about "fixed" action patterns (<u>Erbkoorination</u>)
as they have come to be called (although slightly
erroneously). Second, I shall consider the problem of
sequential variability, or the ways in which motor actions are
coupled together to form chains of on-going behavior.

A. "Fixed" Action Patterns

    There are few, if any, concepts that are more central to
the development of modern ethological ideas than that of the
"fixed" action pattern (for reviews see Schleidt, 1974;
Eibl-Eibesfeldt, 1975; Barlow, 1977). A variety of terms have
been applied to describe the relatively discrete motor patterns that <u>appear</u> "fixed" (Schleidt, 1974; Barlow, 1977),
however, until very recently, there were no systematic
attempts to measure accurately how "fixed" "fixed" action
patterns are (Schleidt, 1974). Among the diagnostic criteria
used to characterize "fixed" action patterns (see above
references for review), two that have been studied in greatest detail within the last decade or so involve statistical
measurement of form and sequencing.

B. A Histological Approach to Behavioral Morphology and
    Variability

    A number of problems surface when one wishes to measure
precisely behavioral morphology, namely, choosing <u>what</u> to
measure and <u>how</u> to do it. The determination of "the" most
biologically important component (e.g., the parameter of a
signal that carries the "most important" information) often is
very difficult, if not impossible. And, of course, the way
in which various components of a motor pattern interact with
one another (covary, see for example Stokes, 1962; Golani,
1973, 1976; Wiley, 1975) must be dealt with also. The "size"
of a movement as well as its duration, rate, and patterning
(e.g., in terms of inter-act intervals) may all contribute to
its function in various contexts, although each component may
vary independently (Stamps and Barlow, 1973; Wiley, 1975).
Therefore, using only one criterion may be misleading (Stamps
and Barlow, 1973; Schleidt, 1974; Barlow, 1977). Nonetheless,
if research is to commence, it may not be bad practice to
begin to measure the characteristics of a motor act(s) that
"appear" to the eye (or ear) to be the least variable,
realizing, of course that other components and seemingly more
variable acts must also be dealt with. It would be misleading
to concentrate solely on attempting to "prove" stereotypy of

"fixity", without dealing with patent variability.

After determining what to measure, the problem of how to do it must be confronted. For some behavioral phenotypes such as sounds, sonographic analysis can provide an accurate picture. Fossils also provide good "subjects" for measurements of behavioral variability (e.g., Berg and Nishenko, 1975). On the other hand, visual displays and visible action are more difficult to analyze. However, a number of investigators have solved the problem in a variety of ways through the use of video equipment and laudatory perseverance (e.g., Dane, Walcott, and Drury, 1959; Baerends and van der Cingel, 1962; Dane and van der Kloot, 1964; van Tets, 1965; Hazlett, 1972a,b; Golani, 1973, 1976; Wiley, 1973, 1975; Crews, 1975; Jennsen, 1975; Johnston, 1976; Pitcairn and Schleidt, 1976; Hill and Bekoff, In Press). Some problems that have surfaced in the use of video equipment to analyze visible motor actions include changes (however slight) in film speed (Stamps and Barlow, 1973; Wiley, 1973; Crews, 1975; Bekoff, unpub. data) and orientation, both of which must be dealt with in order to produce accurate and reliable measurements.

Assuming that the measurement can be made with a high degree of accuracy, the appropriate statistical method must be used. Although there are a number of statistics that can be used to measure variability, the coefficient of variation (CV) provides a direct measure and has been the statistic used in the majority of cases in which variability has actually been assessed. The CV is simply the standard deviation expressed as a percentage of the mean (CV = SD x $100/\overline{X}$; Schleidt (1974) does not multiply SD by 100). One could also measure stereotypy by taking the reciprocal of the CV. Barlow (1977) has presented a measure of stereotypy (ST) to account for the rare instance(s) when CV = 0 and consequently ST = infinity. He suggests adding 1 to the denominator of the reciprocal of the CV and derives the following equation

$$ST = \overline{X}/SD + .01\overline{X}$$

When CV = 0, ST = 100, when CV = 1, ST = 50, and when CV = 100, ST = 0.99, etc. Since virtually all published studies have reported measures of variation, I shall concentrate on CV values. Furthermore, since variability and not invariant fixity or stereotypy seems to be the case in all published examples, it seems wiser to concentrate on measuring what is there, namely variability, and not regard it as "slop" in the system (Barlow, 1977) or noise (Klingsporn, 1973), in one's quest for finding fixity. Dawkins and Dawkins (1973, p. 93) present a formula for determining whether or not the difference between two CV's is statistically significant.

Before considering some studies in which variability has

been measured, it is appropriate to discuss briefly whether or not it might be possible to define some "cut-off" point be- tween the categories of "relatively fixed" (modal--see below) and "variable" or "unstereotyped". In a sense, we are asking whether or not it would be possible to set up a standard CV (or ST) against which a variety of data could be compared and against which we might test our own perceptibility. That is, it might be the case that we are able to recognize an act(s) (or sound(s)) and differentiate it from another act(s) when the CV is less than 10% (or 0.1, Schleidt, 1974). On the other hand, we must also remember that our subjects are un- doubtedly better perceivers of their own conspecific behavior than are we, and a 10% variability may be too high. Further- more, "mood" and context can affect variability (see below), and what may be important are relative differences in variation. In the absence of relevant experimentation that could answer questions dealing with motor pattern variation and perception (for the very few exceptions see Barlow, 1977, p. 109 ff), it seems premature to attempt to define a cut-off point between "relative fixed" and "variable". Comparisons between behavioral variability and morphological variation are problematic (Wiley, 1973) and at the moment, further speculation should be replaced by quantitative research.

C. The Morphology of Visible Action

Since the idea of "fixed" action patterns generated a lot of controversy during the 1940s and subsequent decades, it is surprising to find that there have actually been very few detailed quantitative analyses of actions that have been assigned to the category of "fixed". Perhaps a lot more light and considerably less heat would have been produced had rigorous analysis replaced both unfounded supportive en- thusiasm as well as unsubstantiated pessimism and denial. In addition, the distinction between intra-individual and inter- individual variability often was (is) overlooked (as well as being unmeasured). For example, if a signal transmits in- dividual identity, the CV might be expected to be low within an individual but higher (the difference must be detectable) when different individuals are compared to one another (Slater, personal communication).

The first published quantitative attempt to answer questions relating to inter-individual behavioral variability was performed by Dane, Walcott, and Drury (1959); see also Dane and van der Kloot (1964). They studied various displays of Goldeneyes (Bucephala clangula) and used duration as the analyzable phenotype. The rapidity with which many actions occur suggested superficial invariance, and, in fact, some of

the displays of the Goldeneyes are relatively fixed. For
example, the CV for 66 simple head-throws by males was 6.20%
($\bar{X}$ = 1.29 sec., SD = 0.08 sec.). On the other hand, the CV of
13 masthead displays by males was 52.08% ($\bar{X}$ = 4.80 sec.,
SD = 2.50 sec.). The large amount of variation in the masthead display, however, is not due to the difference in the
speed of the movements, but rather, is due to the variable
length of the pauses at the beginning and end of the active
part of the movements themselves. If the pauses are omitted
from the analysis, the CV for the masthead display is equal
to 7.69% ($\bar{X}$ = 0.65 sec., SD = 0.05 sec.). But, not all of the
displays of the male Goldeneye show CV's less than 10.0%.
For example, the bowsprit, nodding and ticking had CV's of
17.14%, 72.22%, and 26.32%, respectively. Female displays for
which data were presented show on the average greater
variability than do displays by males. This first attempt to
quantify a "classical" ethological concept showed clearly that
although some displays showed relative fixity, ideas about
invariant stereotypy were not supported. A more recent
example also points to the fact that finding CV's = 0 is
highly unlikely, but that, indeed, some bird displays are
remarkedly fixed, nonetheless. Wiley (1973) studied the strut
display of male sage grouse (<u>Centrocercus uropasianus</u>). Using
duration as his measure, Wiley found marked stereotypy when
comparing different males (Table 6). The very slight differences that were found showed no consistent relationship with
breeding success. CV's for the time interval between the same
point in successive struts ranged from 9.8% to 34.2%. Inter-

TABLE 6

The Variability of the Duration of Struts
by Individual Male Sage Grouse (from Wiley, 1973)

Male	Coefficient of Variation (%)
D	3.3
UT	1.7
A	3.3, 1.7, 1.7
B	3.4, 2.1
C	1.7
K	4.3
I	2.6

vals varied even when there was little change in the male's
external situation. Marked stereotypy for inter-strut intervals was found. For one male, the CV for the mean interval
for 45 consecutive struts equaled 0.71%. In comparison with
other data that are available concerning the variability of
motor patterns, the strut display of male sage grouse is among
the most stereotyped (with the exception of songbirds -- see
below; for summary tables showing CV's and measures of stereotypy for a variety of species and acts, see Hazlett, 1972a and
Barlow, 1977).

D.   Context as a Factor in Behavioral Variability

     As mentioned above, context appears to be an important
variable in determining the "form" of a particular act, although there are a few studies that have asked the specific
question of how context and behavioral variability are inter-
related. The most stereotyped motor coordinations appear to
be those that are important in locomotion or communication
(Wiley, 1973; Schleidt, 1974). Schleidt (1974) further
suggests that actions that occur at a very high frequency
and/or are adapted to deal with uniform situations or objects
such as pecking, biting, chewing, swallowing, digging, and
preening should also be highly stereotyped. Hazlett (1972b)
studied the variability of movements of the chelipeds and
ambulatory legs of spider crabs (Microphyrs bicornutus) in
different contexts. He found that there was less variability
during agonistic displays than during non-display (e.g.,
feeding (cheliped movements) and walking (ambulatory leg
movements) behaviors). None of his CV's, however, are less
than 10.0% (see his Table 1). Nonetheless, the displays are
clearly different from the non-display behaviors and both
humans and spider crabs are able to tell them apart. Wiley's
study (see above) also showed that display behavior can be
extremely stereotyped. However, there are exceptions to the
rule that behaviors used in a communicative situation are more
stereotyped than non-communicative behaviors. For example,
the hermit crab Petrochirus diogenes shows slightly higher
variability in the movement of the chelipeds during agonistic
"displays" than during walking (see Hazlett, 1972a for further
details).
     Although there are data concerning behavioral variability
available for a number of different animal groups, in searching the literature I could find no equivalent data for
mammals (although the same techniques that are used on other
animals could be employed). We (Hill and Bekoff, In Press)
undertook a study of the variability of some motor acts that
occur during agonistic interactions and social play in

TABLE 7

The variability in the duration of three acts performed by infant Eastern coyotes in two contexts (from Hill and Bekoff, 1977). The high CV values reported in this and the next two tables are most probably due to the distributions being positively skewed. For purposes of comparison between groups, it is important that the distributions be the same. For the groups compared in Tables 7-9, this was the case. For a discussion of different frequency distributions and behavioral variability see Schleidt (1974).

Act	Coefficient of Variation (%)	
	Agonistic Behavior (n=46)	Social Play (n=202)
Stand-over	105.90 (n=46)	80.36* (n=51)
Body-bite	144.83 (n=70)	108.14 (n=186)
Scruff-bite	142.59 (n=47)	120.19 (n=56)

* The differences between stand-overs and body-bites are significant (using the "small c" statistic of Dawkins and Dawkins (1973)).

Eastern coyotes (see above). Briefly, we analyzed the duration of three different acts in order to determine if actions that are used during play are actually more "exaggerated" than the same actions when they appear in non-play situations (for a full discussion of these ideas see Henry and Herrero, 1974; Bekoff, 1976; Hill and Bekoff, In Press). We also calculated CV's (Table 7) to determine variability. Note that the CV's for duration in all three acts in the two different contexts are extremely high (in five out of the six cases, the SD was greater than the mean, hence a CV greater than 100%). With respect to the two different types of biting, analysis of the form of the actions did not appear to differ from context to context. That is, during fighting, it did not appear that the mouth was open any wider than during play. In addition, there was no significant difference in the mean rate of biting during play when compared to agonistic interactions.

    The results of this study demonstrate that there is extreme variability in the duration of the three acts that were performed during social interactions in which there were two (or more) active participants. Obviously, what one individual does will have some effect on the behavior of the other(s). For example, the duration and rate of biting may be affected by the type of interaction (e.g., chase versus close-contact).

E. The Morphology of Sounds

The above examples involve analyses of behavioral patterns that can be observed with the eye. It is also possible to ask questions about variability for auditory behavior. In fact, there have been a number of analyses of bird vocalizations, and although many more data are needed, it appears that bird vocalizations show less variability overall than do visible behaviors. For example, Marler and Isaac (1960) found CV's of 1.3% and 1.7% for the intervals between successive syllables in the songs of two individual chipping sparrows (Spizella passerina). Mulligan (1963) analyzed the duration of seven types of song sung by two song sparrows (Melospiza melodia) and reported CV's ranging from 8-15%. However, Heckenlively (1970) reported CV's for five different aspects of the songs of four different birds and only three of 18 CV's were less than 10%. For more data on bird sounds see Schleidt (1974), Wilkinson and Howse (1976), and Wiley (1976).

In conclusion of this section, a number of points should be made. First, and foremost, there appear to be extremely few, if any, behavioral patterns that are absolutely fixed (e.g., CV = 0 or even less than 1.0%). Indeed, the implication of absolute morphological rigidity in the translation of Erbkoordination was unfortunate, and according to Eibl-Eibesfeldt (1975) was not originally intended. Barlow (1968) has suggested that it might be wiser and probably more correct to drop the word "fixed" and substitute the word "modal" (to refer to the most frequently occurring "form" of a given act), since the spatiotemporal patterns of coordinated movement usually cluster around some mode (Barlow, 1968, p. 230). Quantitative data support Barlow's suggestion (see also Barlow, 1977). In addition to terminological problems, it is important to stress that the "meaning" of the value of CV's will undoubtedly vary from species to species and from act to act. Nonetheless, the CV (or ST) provides a good quantitative measure of behavioral variability and hopefully future studies of motor pattern variability will contain detailed analyses of pertinent data. Consideration of context must also be included. The decision of what to measure also deserves careful attention, e.g., Hazlett's (1966) use of models to establish that limb position was important before measurements were done on variability of limb position (Hazlett, 1972a). Singling out duration (or other temporal factors) is perhaps a convenient and useful way to begin, however, the final analysis will of necessity have to account for the different components

that make up the act (i.e., spatial characteristics) as well
as their sequencing (see below). Finally, analyses of "fixed"
action patterns lend themselves to a multidisciplinary approach that will certainly provide pertinent data concerning
the neuromuscular bases for the generation of these motor
patterns, and motor behavior in general. The reader is referred to the fine work that has been conducted on the escape
response of the Pacific nudibranch mollusc, Tritonia diomedia
(Willows and Hoyle, 1969; Willows, Dorsett, and Hoyle,
1973a,b; Hoyle and Willows, 1973; Dorsett, Willows, and Hoyle,
1973; Hoyle, 1975).

F. Behavioral Sequences

      An analysis of the variability of individual motor acts
does not provide any information concerning the ways in which
these behaviors, stereotyped or not, are linked together to
form continuous chains of behavior. As McFarland (1976; see
also Sibly and McFarland, 1976) has pointed out, selection
can operate not only on species-typical characteristics of
behavior patterns but also on the order in which they are
performed. McKinney (1961) as well has suggested that behavioral sequences can serve display function. Sequence analysis
has become a popular method nowadays. It may be used to
analyze sequential dependencies or transitions between two or
more acts for a wide variety of behavioral patterns including
aggression, grooming, courtship, and play, and also has been
used to study communication networks in animal groups. Its
wide use has continued despite some obvious drawbacks in
available techniques (Slater, 1973; see below). In this
paper, I shall be concerned particularly with the question of
how "fixed" are behavioral sequences and how may sequential
variability be studied. One of the diagnostic criterion for
"fixed" action patterns concerns the temporal order in which
actions comprising a sequence occur (Reese, 1963; Barlow,
1968, 1977; Stamps and Barlow, 1973; Schleidt, 1974).
      My own interests in behavioral sequencing stemmed from
earlier work on behavioral development in canids (Bekoff,
1972, 1974, 1977, In Press[a]). Questions to which I was seek-
ing answers included (1) does the variability of sequencing
change during ontogeny and (2) do behavioral sequences performed in different contexts differ with respect to their
variability? In an attempt to learn about the methods available for doing such analyses I read a variety of research
papers and review articles (e.g., Nelson, 1964,1973; Altmann,
1965; Hazlett and Bossert, 1965; Delius, 1969; Chatfield and
Lemon, 1970; Slater and Ollason, 1972; Slater, 1973, 1975;
Chatfield, 1973; Metz, 1974; Dawkins, 1976; Morgan, Simpson,

Hanby, and Hall-Craggs, 1976) and texts (e.g., Suppes and Atkinson, 1960; Maxwell, 1961; Karlin and Taylor, 1975). To make a long story short, it became (too) obvious to me that there actually was little agreement among students of behavior about what techniques were appropriate and which were inappropriate for use on behavioral data specifically collected during social interactions between two (or more) individuals. There seemed to be two groups, the "over-killers" and the conservatives. The former group went head-on into some rather sophisticated mathematical procedures either ignoring some very basic underlying assumptions or recognizing their existence and continuing nonetheless. The latter group took hold of the situation, recognized that their questions could be answered using other (standard) procedures, and proceeded to use the techniques that would require the least "molding" of their animals (and research interests) into something that they were not. It seemed to me that most of the animals that had been studied had a rather long phylogenetic history and that it was presumptuous of a modern-day researcher to try to change history with the stroke of a pen or the use of a computer! The best course to follow appears to be to assume as little as possible and not destroy the "integrity" of your animals in order to fit certain underlying ground rules such as stationarity (see below). In a very timely and interesting paper, van der Kloot and Morse (1975) suggest that at first, one should assume things that they recognize to be unlikely just to get started, and then as the analysis proceeds, to be "... vigilant about the possibility that apparently significant points are actually owing to non-stationarity" (p. 182). While this may, in fact, "get the ball rolling", I believe that time may be wasted by following this procedure and also that "turning back" may be difficult. As Rensch (1971, p. 26) has written:

> Sometimes, however, biological research has been dazzled by the very precision of mathematical methods. The formal correctness of a calculation, it must be remembered, is only of value if it rests on adequate material correctness; that is to say, if the premises for applying particular formulas and mathematical proofs are appropriate.

G. Some General Comments on Sequence Analysis and the Stationarity Problem

A very common analysis of sequence data involves the use of transition, or conditional, probabilities. Such Markovian analyses have been performed on a wide variety of animals (Altmann, 1965; Chatfield and Lemon, 1970; Slater, 1973;

Metz, 1974). However, because of the underlying assumptions that govern this type of analysis, behavioral data only rarely meet the requirements for use of this method. The attractive simplicity of Markov analyses (and processes) is persuasive in making us overlook complexity (Nelson, 1973). One of the main factors that limits the application of Markovian analyses to behavioral data is the requirement of stationarity.[1] A stationary sequence is one in which the probability structure does not change with time (regardless of the starting point; Pierce, 1961, p. 58), or in animal behavior research, it would be necessary to demonstrate that the animal is in a steady-state (Slater, 1973). Of course, animals are not (usually) steady-state systems, so the question becomes one of grade. A number of researchers have made it very clear that Markovian analyses require adherence to the assumption of stationarity (Lemon and Chatfield, 1970; Fernald, 1973; Schleidt, 1974; Slater, 1973, 1975; Still, 1976; Wickler, 1976) and the lack of ability to adhere to this prerequisite is the main limiting factor in comparing a probability chain with a random model (Slater, 1973) and making further statistical inferences (Chatfield, 1973; also see Oden, this volume; for a complete discussion of stationarity see Karlin and Taylor, 1975, chapter 9). Some investigators (Nelson, 1964; Lemon and Chatfield, 1971; Heiligenberg, 1973; Baylis, 1975; Verberne and Leyhausen, 1976) have done tests for stationarity by analyzing data from different segments of one observation period or from a number of observation periods over several weeks. Such tests of internal consistency are essential! In addition, it should be stressed that even under steady-state or constant external conditions, there may be fluctuations in the readiness of an individual animal to perform a particular behavior (Heiligenberg, 1973, 1976). (Hazlett and Estabrook's (1974) method of looking at narrowly defined portions of fights avoids some of the problems raised by the requirement for stationarity.) Therefore, maintaining constant external conditions in an experimental design does not in any way guarantee stationarity for the "internal" state of the animal.

Along these lines, it is important to mention briefly that the nature of the conditions under which most observations of social behavior are conducted, almost, in themselves, preclude the ability to satisfy the assumption of stationarity. Some examples include: (1) when two or more

---

1. Editor's footnote: Some types of Markovian processes do not show absolute stationarity, but cyclic, regular changes in probability. See N. T. J. Baily's The Elements of Stochastic Processes with Applications to the Natural Sciences, 1964, John Wiley & Sons.

individuals are interacting;   (2) when data for different individuals are lumped (see Chatfield, 1973);   (3) developmental studies, during which individuals are forming social
relations and perceptual-motor skills are being acquired and
"polished";   (4) motivational studies (As Slater (1973, p.
145) has written, assuming a steady-state "... is tantamount
to ignoring the possibility that motivational changes
occur.");   (5) when signals have a cumulative "tonic" effect
(Schleidt, 1973).   With respect to point 3, any study in
which relationships are forming would probably preclude
assuming stationarity (Fernald, 1973; p. 342).   Therefore,
what with the extreme constraints on satisfying the assumptions of stationarity, it is not at all surprising to find
that it is the nature of non-stationarity that becomes of
interest (Fernald, 1973; Slater, 1975; Wickler, 1976)!   As
Slater (1975, p. 12) has noted, "Non-stationarity, the
bug-bear of the sequence analyst, is thus the bread and butter
of anyone interested in motivation "and/or other behavioral
changes
     I have gone through the above discussion in some length
because of the fact that although "warning flags" have been
waved countless times by ethologists, statisticians, and
mathematicians, alike, the application of Markovian analyses
continues rather freely.   Furthermore, it is not quite the
case that questions about behavioral variability can be
answered directly by Markovian analyses anyway.

H.   Overcoming the "Bug-bear" of Stationarity

     Given the fact that certain mathematical procedures
require assumptions that have a very low probability of being
met in behavioral research, one is now faced with the problem
of what to do with his or her data if a question about
behavioral variability is at hand.   There are a number of alternatives.   First, the researcher can choose another problem!   Second, he or she can ignore either intentionally or
unintentionally, the underlying ground rules.   Third, and
most appealing, the investigator can find other ways to
answer the questions of interest.   With respect to the organization of behavioral patterns throughout a sequence, and not
necessarily with transitions between individual acts, one
technique that has been used is a method that I call Action
Unit Analysis (see van Iersel and Bol, 1958; Ainley, 1974;
Bekoff, Ainley, and Bekoff, In Rev.)   This technique is useful
in providing information about the distribution of acts
throughout a sequence(s) but does not provide any information
about the way in which acts are linked with one another.
There are also a number of ways in which single cells in

transition matrices can be analyzed to detect statistically
significant transitions (Andrew, 1956b; Ainley, 1974; Poole
and Fish, 1975, 1976; Stevenson and Poole, 1976; Bekoff,
Ainley, and Bekoff, In Rev.; for a review of a variety of
procedures see Fagen, unpub. ms.). Since the main interest
at hand concerns behavioral variability, let me briefly discuss some examples in which behavioral sequences have been
analyzed with this question in mind.

I. Sequential Variability

If one casts his or her data into an n x m transition
matrix, a chi-square analysis can be performed to give some
very general information about whether or not the acts are
randomly distributed. One can also perform Markov and information theoretic analyses but these procedures depend on
assumptions that are very difficult to satisfy. An alternative approach that has been used only in a few instances is
to study variability within behavioral sequences using the
coefficient of variation as the statistical measure (Dawkins
and Dawkins, 1973; Baker, 1973). Both the works of the
Dawkins and Baker, however, did not involve analyses of
sequential variability, but rather were concerned with the
duration of acts (and intervals) within sequences (Dawkins and
Dawkins, 1973, Table 3; Baker, 1973, Table 1).
The CV can also be used to assess sequential variability.
We have applied this technique to two different studies. For
both, raw data were cast into transition matrices and conditional probabilities for two-act transitions were calculated
for each of the cells. Then, means and standard deviations
for selected transition probabilities within each matrix were
calculated (see Bekoff, 1977; Bekoff, Ainley, and Bekoff,
In Review) and CV's were generated for each table, in turn.
In the first analysis, we were interested in testing the
hypothesis (see Bekoff, 1976 for details) that social play
behavior is more variable than non-play behavior. Forty-
eight sequences of social play and 70 sequences of agonistic
behavior were analyzed for a pair of infant coyotes. The
variability in the coupling of two-act transitions for inter-
individual sequences was assessed. The results are presented
in Table 8. It can be seen that the variability in the
coupling of two-act transitions is more variable in social
play than during agonistic interactions (c = 6.28, df = 882,
p < 0.001; see Dawkins and Dawkins, 1973, p. 93 for formula
for "c"). It should be mentioned that analysis of the same
data using information theoretic measures, namely, conditional
uncertainties, yielded the same results (R. Fagen, pers.
comm.). We chose not to use conditional uncertainties because

TABLE 8

The Variability in the Coupling
of Two-act Transitions by the Same Pair of
Coyotes in Two Social Contexts (see Table 7)

Coefficient of Variation (%)	
Social Play	Agonistic Behavior
(n=445)*	(n=439)
98.02	72.27

* The number of two-act transitions (cells) for
individual behaviors that occurred more than
30 times.

we could not assume stationarity with any confidence. Day-
to-day changes in sequences are obvious (Bekoff, unpub. data).

J.   Behavioral Ontogeny and Sequential Variability

     The relationship between ontogeny and behavioral varia-
bility has been little studied. Chatfield and Lemon (1970)
suggest that perhaps the order of dependency between acts increases with age. More importantly, they stress that the
techniques with which their paper was concerned (chi-square
goodness-of-fit tests for Markov chains and information
theoretic analyses) should only be used if the population is
homogenous. There are some data that indicate that the behavior of young animals is not necessarily more variable than
that of adults (Schleidt and Shalter, 1973; Wiley, 1973).
And, one very interesting study on the ontogeny of predatory
behavior in marine snails (Berg, 1976) has demonstrated
increasing variability as the animals age and gain more experience. Of course, more data are needed before any general
statement (if indeed this will even be possible) can be made
concerning the relationship between behavioral ontogeny and
variability. Two important factors to consider will be the
state of the organism at birth (altricial or precocial) and
the modifiability of the nervous system during postnatal
life.
     With respect to the variability of sequences of behavior,
I could find no published study in which these types of data
have been reported. We studied the development and organization of comfort behavior in Adelie penguins (Pygoscelis
adeliae) at the Cape Crozier Rooker, Ross Island, Antarctica

TABLE 9

The Variability in the Coupling of Two-act Transitions During Comfort Behavior by Adelie Penguins of Different Ages (see Bekoff, 1977b and Bekoff, Ainley, and Bekoff, In Review for details)

	Group	Coefficient of Variation (%)
Non-oil	7-13 days (n=9)*	90.10
Chicks	14-20 days (n=50)	84.16
	21-28 days (n=25)	76.59
	Non-oil adults (n=88)	96.66
Dry-oil	Chicks (n=89)	109.55
	Adults (n=132)	143.76
Wet-oil**	Adults (n=130)	135.05

* The number of two-act transitions (cells) for individual behaviors that occurred more than 15 times.
** Calculated from Ainley (1970).

(Bekoff, Ainley, and Bekoff, In Review). Coefficients of variation were calculated for selected two-act transitions for the different groups of birds during different types of comfort sequences (Table 9). There were no significant differences between chicks and adults when non-oil sequences were performed (there were not enough observations for the youngest group of chicks). Dry-oiling by chicks was significantly less variable than either dry-oiling or wet-oiling by adults. These results for non-oiling agree with the suggestions of Schleidt and Shalter (1973) and Wiley (1973), namely, that the behavior of young animals is not necessarily more variable than that of adults. On the other hand, the results for dry-oiling agree with Berg's (1976) results, namely, that there is increasing variability with age. The implications of these results and discussions of behavioral variability and ontogeny, in general, are considered in Bekoff (1977b).

The two above examples show that sequential variability between two-act transitions can be analyzed using a standard statistical procedure. Furthermore, it should be emphasized that Markov chain analyses are difficult to apply to a vast majority of behavioral studies and that they do not have much bearing on problems of sequential variability.

CONCLUSION

The "classical" ethological concepts that have been discussed in this paper are still very much with us today. Techniques for assessing social dominance hierarchies and for performing detailed behavioral taxonomic studies are rather straightforward. They are readily available and should be used. Also, quantitative analyses of the variability of motor patterns using the coefficient of variation are easy to do (actually measuring the component of interest may be difficult) and should be included in all discussion of "fixed" or "modal" action patterns. The analysis of behavior sequences presents another story. The inability of validly assuming stationarity is a major limitation to the application of standard Markovian analyses. The ability of an animal to change "states" and to perform patterns of behavior that vary with time are probably the very reasons why so many of us study behavior in the first place! Temporal changes that make the assumption of stationarity a rather questionable procedure appear to be the rule, rather than the exception. Our methods must reflect this fact. While I fully agree with Dingle (1972, p. 149) that "... whole areas of ethology, with their concomitant theories, have built up virtually in the absence of quantitative studies...", it must be emphasized that misuse of quantitative procedures will certainly, in the end, have more damaging than beneficial results.

A great need in behavioral science today is for the development of models and statistical procedures that take into account the individual idiosyncracies of the behavioral systems to which they are going to be applied. Researchers must go out and look at their animals and fit a method of analysis to their subjects, not vice versa. Using techniques that require our making assumptions that destroy the distinctiveness of the species, the individuals, or the problem under study must be avoided. Mathematical convenience must be replaced with behavioral realities. As Dobzhansky (1973, p. 59) has written, "... the complexities of the order of nature should not be evaded...the only way to simplify nature is to study it as it is, not as we would have liked it to be".

ACKNOWLEDGEMENTS

I would like to thank the following people for discussing many of the issues contained herein: Wayne Aspey, George Barlow, Jeffrey Baylis, Benjamin Beck, Ivan Chase, Debra Christein, William Corning, Victor DeGhett, Robert Fagen, Charles Fuenzalida, Randall Lockwood, and Jeffry Mitton.

Peter Slater, Harriet Hill, Wayne Aspey, and Ivan Chase provided useful comments on an earlier version of this chapter.
All errors of omission and commission are mine. Harriet Hill,
Robert Jamieson, Susanne King, and Judy Diamond helped with
animal care and observations. Nutan Pall kindly typed the
manuscript. My own research reported herein was supported in
part by a Faculty Research Initiation Fellowship and Biomedical Support Grant from the University of Colorado.
Travel to the symposium was supported in part by the Graduate
School of the University of Colorado.

REFERENCES

Ainley, D.G. 1970. Communication and reproductive cycles of
     the Adelie penguin. Unpub. Ph.D. dissert., Johns Hopkins
     Univ., Baltimore, Maryland.
_____. 1974. The comfort behavior of Adelie and other pen-
     guins. Behaviour 50: 16-51.
Altmann, J. 1974. Observational study of behavior: Sampling
     methods. Behaviour 48: 227-267.
Altmann, S.A. 1965. Sociobiology of rhesus monkeys. II.
     Stochastics of social communication. J. Theoret. Biol.
     8: 490-522.
Andrew, R.J. 1956a. Intention movements of flight in certain
     passerines and their use in systematics. Behaviour 10:
     179-204.
_____. 1956b. Normal and irrelevant toilet behavior in
     Emberiza spp. Anim. Behav. 4: 85-91.
Atz, J.W. 1970. The application of the idea of homology to
     behavior. p.53-74 in L.R. Aronson, E. Tobach, D.S.
     Lehrman, and J.S. Rosenblatt (eds.) Development and
     Evolution of Behavior, W.H. Freeman & Co., San Francisco.
Baerends, G.P. and N.A. van der Cingel. 1962. On the
     phylogenetic origin of the snap display in the common her-
     on (Ardea cinerea L.). Symp. Zool. Soc. Lond. 8: 7-24.
Baker, M.C. 1973. Stochastic properties of the foraging be-
     havior of six species of migratory shorebirds. Behaviour
     45: 242-270.
Barber, H.S. 1953. North American fireflies of the genus
     Photoris. Smithson. Misc. Coll. 117 (1): 1-58.
Barlow, G.W. 1968. Ethological units of behavior. p. 217-232
     in D. Ingle (ed.) The Central Nervous System and Fish
     Behavior, Univ. of Chicago Press, Chicago.
_____. 1977. Modal action patterns. p. 94-125 in T.A. Sebeok
     (ed.) How Animals Communicate, Univ. of Indiana Press,
     Bloomington.
Baylis, J. 1975. A quantitative analysis of long-term court-
     ship in two sympatric species of cichlid fish. Unpub.

Ph.D. dissert., Univ. of California, Berkeley.

Bekoff, M. 1972. The development of social interaction, play, and metacommunication in mammals: An ethological perspective. Quart. Rev. Biol. 47: 412-434.

_____. 1974. Social play and play-soliciting by infant canids. Amer. Zool. 14: 323-340.

_____. 1975. Social behavior and ecology of the African canidae: A review. p. 120-142 in M.W. Fox (ed.) The Wild Canids, Van Nostrand Reinhold, New York.

_____. 1976. Social play: Problems and perspectives. p. 165-188 in P.P.G. Bateson and P. Klopfer (eds.) Perspectives in Ethology, Vol. 2, Plenum Press, New York.

_____. 1977a. Social development, social bonding, and dispersal in canids: The significance of species and individual differences in behavioral ontogeny. IVth Inter. Cat Conf., Seattle.

_____. 1977b. The ontogeny of behavior in Adelie penguins with some comments on behavioral variability. Paper presented at the mettings of the Animal Behavior Society, Pennsylvania State Univ.

_____. In Press$_a$. Mammalian dispersal and the ontogeny of individual behavioral phenotypes. Amer. Nat.

_____. In Press$_b$. Behavioral development in coyotes and Eastern coyotes. In M. Bekoff (ed.) Coyote Biology: Evolution, Behavior, Ecology, and Management, Academic Press, New York.

_____, D.G. Ainley, and A. Bekoff, In Review. The organization and ontogeny of comfort behavior in Adelie penguins.

_____, H. Hill, and J.B. Mitton. 1975. Behavioral taxonomy in canids by discriminant function analyses. Science 190: 1223-1225.

Bell, C.C., J.P. Myers, and C.J. Russell. 1974. Electric organ discharge patterns during dominance related behavioral displays in Gnathonemus petersii (Mormyridae). J. Comp. Physiol. 92: 201-228.

Berg, C.J. 1974. A comparative ethological analysis of strombid gastropods. Behaviour 51: 274-322.

_____. 1976. Ontogeny of predatory behavior in marine snails (Prosobranchia: Naticidae). The Nautilus 90: 1-4.

_____ and S. Nishenko. 1975. Stereotypy of predatory boring behavior of Pleistocene naticid gastropods. Paleobiol. 1: 258-260.

Bernstein, I.S. 1970. Primate status hierarchies. p. 71-109 in L. Rosenblum (ed.) Primate Behavior, Academic Press, New York.

Brantas, G.C. 1968. On the dominance order in Friesian-Dutch dairy cows. Z. Tierzuch. Zuchtungsbiol. 84: 127-151.

Breed, M.D. 1976. The evolution of social behavior in primitively social bees: A multivariate analysis. Evo-

lution 30: 234-240.
Brown, L.E. and J.R. Brown. 1972. Call types of the Rana
   pipiens complex in Illinois. Science 176: 928-929.
Candland, D.K., D.C. Bryan, B.L. Nazar, K.J. Kopf, and
   M. Sendor. 1970. Squirrel monkey heart rate during for-
   mation of status orders. J. Comp. Physiol. Psychol. 70:
   417-423.
_____, L. Dresdale, J. Leiphart, D. Bryan, C. Johnson, and
   B. Nazar. 1973. Social structure of the squirrel monkey
   (Saimiri sciureus, Iquitos): Relationships among behavior,
   heartrate, and physical distance. Folia Primat. 20:
   211-240.
_____, D.B. Taylor, L. Dresdale, J.M. Leiphart, and S.P.
   Solow. 1969. Heart rate, aggression, and dominance in the
   domestic chicken. J. Comp. Physiol. Psychol. 67: 70-76.
Cane, V.R. 1959. Behavior sequences as semi-Markov chains.
   J. Roy. Stat. Soc. 21B: 36-58.
Cattell, R.B. 1949. r_p and other coefficients of pattern
   similarity. Psychometrika 14: 279-298.
_____, C.R. Bolz, and B. Korth. 1973. Behavioral types in
   purebred dogs objectively determined by taxonome. Behav.
   Genet. 3: 205-216.
Chase, I.D. 1974. Models of hierarchy formation in animal
   societies. Behav. Sci. 19: 374-382.
Chatfield, C. 1973. Statistical inference regarding Markov
   chain models. Appl. Stat. 22: 7-20.
_____, and R.E. Lemon. 1970. Analysing sequences of behavior-
   al events. J. Theoret. Biol. 29: 427-445.
Chercovich, G.M. and S.K. Tatoyan. 1973. Heart rate (radio-
   telemetrical registration) in macaques and baboons according to dominant-submissive rank in a group. Folia
   Primat. 20: 265-273.
Clutton-Brock, T.H., P.J. Greenwood, and R.P. Powell. 1976.
   Ranks and relationships in highland ponies and highland
   cows. Z. Tierpsychol. 41; 202-216.
_____, and P.H. Harvey. 1976. Evolutionary rules and primate
   societies. p.195-237 in P.P.G. Bateson and R.A. Hinde
   (eds.) Growing Points in Ethology, Cambridge Univ. Press,
   New York.
Crews, D. 1975. Inter- and intraindividual variation in dis-
   play patterns in the lizard, Anolis carolinensis.
   Herpetologica 31: 37-47.
Cullen, J.M. 1959. Behaviour as a help in taxonomy. Syst.
   Assoc. Pub. No. 3, 131-140.
Dane, B. and W.G. van der Kloot. 1964. An analysis of the
   display of the goldeneye duck (Bucephala clangula (L)).
   Behaviour 22: 282-382.
_____, C. Walcott, and W.H. Drury. 1959. The form and dura-
   tion of the display actions of the goldeneye (Bucephala

clangula (L)): Behaviour 14: 265-281.

David, H.A. 1959. Tournaments and paired comparison. Biometrika 46: 139-149.

Dawkins, R. 1976. Hierarchical organisation: A candidate principle for ethology. p. 7-54 in P.P.G. Bateson and R. A. Hinde (eds.) Growing Points in Ethology, Cambridge Univ. Press, New York.

_____ and M. Dawkins. 1973. Decisions and the uncertainty of behaviour. Behaviour 45: 83-103.

DeFries, J.C. and G.E. McClearn. 1970. Social dominance and Darwinian fitness in the laboratory mouse. Amer. Nat. 104: 408-411.

Delacour, J. and E. Mayr. 1945. The family Anatidae. Wilson Bull. 57: 3-55.

Delius, J.D. 1969. A stochastic analysis of the maintenance behaviour of skylarks. Behaviour 33: 137-178.

Dewsbury, D.A. 1972. Patterns of copulatory behavior in male mammals. Quart. Rev. Biol. 47: 1-33.

Dilger, W.C. 1960. The comparative ethology of the African parrot genus Agapornis. Z. Tierpsychol. 17: 649-685.

_____ 1962. The behavior of lovebirds. Sci. Amer. 206: 88-98.

Dingle, H. 1972. Aggressive behavior in stomatopods and the use of information theory in the analysis of animal communication. p. 126-156 in H.E. Winn and B.I. Olla (eds.) Behavior of Marine Animals, Vol. 1. Plenum Press, New York.

Dixson, A.F., D.M. Scruton, and J. Herbert. 1975. Behaviour of the Talapoin monkey (Miopithecus talapoin) studied in groups, in the laboratory. J. Zool. Lond. 176: 177-210.

Dobzhansky, T. 1973. Genetic Diversity and Human Equality. Basic Books, New York.

Dorsett, D.A., A.O.D. Willows, and G. Hoyle. 1973. The neuronal basis of behavior in Tritonia. IV. The central origin of a fixed action pattern demonstrated in the isolated brain. J. Neurobiol. 4: 287-300.

Drews, D.R. 1973. Group formation in captive Gelago crassicaudatus: Notes on the dominance concept. Z. Tierpsychol. 32: 425-435.

Dudzinski, M.L., and J.M. Norris. 1970. Principal components analysis as an aid for studying animal behaviour. Forma et Functio 2: 101-109.

Dunbar, R.I.M. 1976. Some aspects of research design and their implications in the observational study of behaviour. Behaviour 58: 78-98.

_____, and J.H. Crook. 1975. Aggression and dominance in the weaver bird, Quelea quelea. Anim. Behav. 23: 450-459.

Dunford, C. and R. Davis 1975. Cliff chipmunk vocalizations and their relevance to the taxonomy of coastal sonoran chipmunks. J. Mammal. 56: 207-212.

Echelle, A.A., A.F. Echelle, and H.S. Fitch. 1971. A comparative analysis of aggressive display in nine species of Costa Rican Anolis. Herpetologica 27: 271-288.

Eibl-Eibesfeldt, I. 1975. Ethology. Holt, Rinehart, & Winston, New York.

Fabre, J.H. 1916. The Hunting Wasps. Hodder & Stoughton, London.

Fagen, R.M. unpub. ms. Behavioural transitions: Cell-by-cell tests for statistical significance.

Fernald, R.D. 1973. A group of barbets: II. Quantitative measures. Z. Tierpsychol. 33: 341-351.

Fox, M.W., and M. Bekoff. 1975. The behaviour of dogs. p. 370-409 in E.S.E. Hafez (ed.) The Behaviour of Domestic Animals, 3rd edit., Bailliere Tindall, London.

Gartlan, J.S. 1964. Dominance in East African monkeys. Proc. East African Acad. 2: 75-79.

_____. 1968. Structure and function in primate society. Folia Primat. 8: 89-120.

Golani, I. 1973. Non-metric analysis of behavioral interaction sequences in captive jackals (Canis sureus). Behaviour 44: 89-112.

_____. 1976. Homeostatic motor processes in mammalian interactions: A choreography of display. p. 69-134 in P.P.G. Bateson and P. Klopfer (eds.) Perspectives in Ethology, Vol. 2, Plenum Press, New York.

Hanby, J. 1976. Sociosexual development in primates. p. 1-67 in ibid.

Harary, F., R.Z. Norman and D. Cartwright. 1965. Structural Models: An Introduction to the Theory of Directed Graphs. John Wiley & Sons, New York.

Hausfater, G. 1975. Dominance and reproduction in baboons (Papio cynocephalus) Contrib. Primat. No. 7, 150 p.

Hazlett, B.A. 1966. Social behavior of the Diogenidae and Paguridae of Curacao. Studies Fauna Curacao 23: 1-143.

_____. 1972a. Ritualization in marine crustacea. p. 97-125 in H.E. Winn and B.L. Olla (eds.) Behavior of Marine Animals, Vol. 1, Plenum Press, New York.

_____. 1972b. Stereotypy of agonistic movements in the spider crab Microphrys bicornutus. Behaviour 42: 270-278.

_____ and W.H. Bossert. 1965. A statistical analysis of the aggressive communications systems of some hermit crabs. Anim. Behav. 13: 357-373.

_____ and G. Estabrook. 1974. Examination of agonistic behavior by character analysis. II. Hermit crabs. Behaviour 49: 88-110.

Heckenlively, D.B. 1970. Song in a population of black-throated sparrows. Condor 72: 24-36.

Heiligenberg, W. 1973. Random processes describing the

occurrence of behavioral patterns in a cichlid fish.
Anim. Behav. 21: 169-182.

_____. 1976. A probabilistic approach to the motivation of
behavior. p. 301-313 in J.C. Fentress (ed.). Simpler Networks of Behavior, Sinauer Assoc., Inc., Sunderland, MA.

Heinroth, O. 1910. Beitrage zur biologie, insbesondere
psychologie und ethologie der Anatiden. Verh. 5th
Internat'l. Ornith. Kongr, 589-702.

Heymer, A. 1975. Der stammesgeschichtliche Aussagewert des
Verhaltens der Libelle Epallage fatime Charp. 1840. Z.
Tierpsychol. 37: 163-181.

Hill, H.L. and M. Bekoff, In Press. The variability of some
motor components of social play and agonistic behaviour in
infant eastern coyotes, Canis latrans var. Anim. Behav.

Hoyle, G. 1975. Identified neurons and the future of neuroethology. J. Exp. Zool 194: 51-74.

_____, and A.O.D. Willows. 1973. Neuronal basis of behavior
in Tritonia. II. Relationship of muscular contraction to
nerve impulse pattern. J. Neurobiol. 4: 239-254.

Hutt, S.J. and C. Hutt. 1970. Direct observation and measurement of behavior. C.C. Thomas, Springfield, Ill.

van Iersel, J.J.A. and A.C.A. Bol. 1958. Preening by two tern
species: A study on displacement behaviour. Behaviour
13: 1-88.

Jenssen, T.A. 1975. Display repertoire of a male
Phenacosaurus heterodermus (Sauria: Iguanidae). Herpetologica 31: 48-55.

Johnsgard, P.A. 1961. The taxonomy of the anatidae: A
behavioural analysis. Ibis 103: 71-85.

Johnston, R.E. 1976. The role of dark chest patches and up-
right postures in the agonistic behavior of male hamsters,
Mesocricetus auratus. Behav. Biol. 17: 161-176.

Karlin, S. and H.M. Taylor. 1975. A First Course in Stochastic Processes, 2nd ed. Academic Press, New York.

Kaston, B.J. 1964. The evolution of spider webs. Amer.
Zool. 4: 191-207.

Kleiman, D.G. 1967. Some aspects of social behavior in the
Canidae. Amer. Zool. 7: 365-372.

Klingsporn, M.J. 1973. The significance of variability.
Behav. Sci. 18: 441-447.

van der Kloot, W. and M.J. Morse. 1975. A stochastic analysis
of the display behavior of the red-breasted merganser
(Mergus serrator). Behaviour 54: 181-216.

van Kreveld, D. 1970. A selective review of dominance-
subordination relations in animals. Gen. Psychol. Mono.
81: 143-173.

Kummer, H. 1973. Dominance versus possession. Symp. IVth
Internat'l. Congr. Primat. 1: 226-231.

Landau, H.G. 1951a. On dominance relations and the structure of animal societies: I. Effect of inherent characteristics. Bull. Math. Biophys. 13: 1-19.

_____. 1951b. On dominance relations and the structure of animal societies: II. Some effects of possible social factors. Bull. Math. Biophysics 13: 245-262.

_____. 1953. On dominance relations and the structure of animal societies: III. The condition for a score structure. Bull. Math. Biophysics 15: 143-148.

_____. 1965. Development of structure in a society when new members are added successively. Bull. Math. Biophysics 27: 151-160.

Lawrence, B. and W.H. Bossert. 1969. The cranial evidence for hybridization in New England Canis. Breviora 330: 1-13.

_____ and _____. 1975. Relationships of North American Canis shown by a multiple character analysis of selected populations. p. 73-86 in M.W. Fox (ed.) The Wild Canids, van Nostrand Reinhold, New York.

LeBoeuf, B.J. 1974. Male-male competition and reproductive success in elephant seals. Amer. Zool. 14: 163-176.

Lemon, R.E. and C. Chatfield. 1971. Organization of song in cardinals. Anim. Behav. 19: 1-17.

Littlejohn, M.J. and R.S. Oldham. 1968. Rana pipiens complex: Mating call structure and taxonomy. Science 163: 1003-1004.

Lockwood, R. 1976. An ethological analysis of social structure and affiliation in captive wolves (Canis lupus). Unpub. Ph.D. dissert., Washington Univ, St. Louis, MO., 362 p.

Lomax, A. 1968. Folk Song Style and Culture. Amer. Assoc. Adv. Sci., Washington, D.C.

_____ with N. Berkowitz. 1972. The evolutionary taxonomy of culture. Science 177: 228-239.

Lorenz, K. 1941. Vergleichende Bewegungsstudien an Anatinen. J. für Ornithol. 89: 194-294.

_____. 1958. The evolution of behavior. Sci. Amer. 199: 67-78.

_____. 1971. Studies in Animal and Human Behaviour. Harvard Univ. Press, Cambridge, Mass.

Manogue, K.R., A.I. Lesher, and D.K. Candlanc. 1975. Dominance status and adrenocortical reactivity to stress in squirrel monkeys (Saimiri sciureus). Primates 16: 457-463.

Marler, P. 1957. Specific distinctiveness in the communication signals of birds. Behaviour 11: 13-39.

_____. 1976. On animal aggression: The roles of strangeness and familiarity. Amer. Psychol. 31: 239-246.

_____ and D. Isaac. 1960. Physical analysis of a simple bird song as exemplified by the chipping sparrow. Condor 62: 124-135.

Maxwell, A.E. 1961. Analysing Qualitative Data. Methuen, London.

Mayr, E. 1958. Behavior and systematics. p. 341-362 in A.

Rose and G.G. Simpson (eds.) Behavior and Evolution, Yale Univ. Press, New Haven, Conn.

McBride, G. 1964. A general theory of social organization and behaviour. Queensland Univ., Brisbane, Fac. Vet. Sci. Papers 1: 75-110.

McFarland, D.J. 1976. Form and function in the temporal organization of behaviour. p. 55-93 in P.P.G. Bateson and R.A. Hinde (eds.) Growing Points in Ethology, Cambridge Univ. Press, New York.

McKinney, F. 1961. An analysis of the displays of the European Eider Somateria mollissima mollissima (Linnaeus) and the Pacific Eider Somateria mollissima v. nigra Bonaparte. Behaviour Suppl. 7.

_____. 1965. The comfort movements of Anatidae. Behaviour 25: 120-220.

Metz, H. 1974. Stochastic models for the temporal fine structure of behaviour sequences. p. 5-86 in D.J. McFarland (ed.) Motivational Control Systems Analysis, Academic Press, New York.

Michener, C.D. 1974. The Social Behavior of the Bees. Harvard Univ. Press, Cambridge, Mass.

Mihok, S. 1976. Behaviour of subarctic red-backed voles (Clethrionomys gapperi athabascae). Canad. J. Zool. 54: 1932-1945.

Moon, J.W. 1968. Topics on Tournaments. Holt, Rinehart, & Winston, New York.

Morgan, B.J.T., M.J.A. Simpson, J.P. Hanby, and J. Hall-Craggs. 1976. Visualizing interaction and sequential data in animal behaviour: Theory and application of cluster-analysis methods. Behaviour 56: 1-43.

Mulligan, J.A. 1963. A description of song sparrow song based on instrumental analysis. Proc. XIIIth Internat'l. Congr. Ornithol., 272-284.

Myers, J.H. and C.J. Krebs. 1971. Genetic, behavioral and reproductive attributes of dispersing field voles Microtus pennsylvanicus and Microtus ochrogaster. Ecol. Mono. 41: 53-78.

_____ and _____. 1974. Population cycles in rodents. Sci. Amer. 230: 38-46.

Nelson, K. 1964. The temporal pattern of courtship behaviour in the glandulocaudine fishes (Ostariophysi, Characidae). Behaviour 24: 90-146.

_____. 1973. Does the holistic study of behavior have a future? p. 281-328 in P.P.G. Bateson and P. Klopfer (eds.) Perspectives in Ethology, Vol. 1, Plenum Press, New York.

Nyby, J., G.A. Dizinno, and G. Whitney. 1976. Social status and ultrasonic vocalizations of male mice. Behav. Biol. 18: 285-289.

Petrunkevich, A. 1926. The value of instinct as a taxonomic

character in spiders. Biol. Bull. 50: 427-432.

Pierce, J.R. 1961. Symbols, Signals, and Noise: The Nature and Process of Communication. Harper and Row, New York.

Pitcairn, T.K. and M. Schleidt. 1976. Dance and decision: An analysis of a courtship dance of the Medlpa, New Guinea. Behaviour 58: 298-316.

Poole, T.B. and J. Fish. 1975. An investigation of playful behaviour in Rattus norvegicus and Mus musculus. J. Zool. Lond. 175: 61-71.

_____ and J. Fish. 1976. An investigation of individual, age and sexual differences in the play of Rattus norvegicus (Mammalia: Rodentia). J. Zool. Lond. 179: 249-260.

Potter, D.A., D.L. Wrensch, and D.E. Johnston. 1976. Aggression and mating success in male spiders. Science 193: 160-161.

Reese, E. 1963. The behavioural mechanisms underlying shell selection by hermit crabs. Behaviour 21: 78-126.

Rensch, B. 1971. Biophilosophy. Columbia Univ. Press, New York.

Richards, S.M. 1974. The concept of dominance and methods for assessment. Anim. Behav. 22: 914-930.

Rovner, J.S. 1973. Copulatory pattern supports generic placement of Schizocosa avida (Walckenaer) (Aranea: Lycosideae). Psyche 80: 245-248.

Rowell, T.E. 1966. Hierarchy in the organization of a captive baboon group. Anim. Behav. 14: 430-443.

_____. 1974. The concept of social dominance. Behav. Biol. 11: 131-154.

Salthe, S.N. 1967. Courtship patterns and the phylogeny of the urodeles. Copeia 1967: 100-117.

Schein, M.W. (ed.) 1975. Social Hierarchy and Dominance. Dowden, Hutchinson, & Ross, Inc., Stroudsburg, Pa.

Schjelderup-Ebbe, T. 1922. Beiträge zur sozialpsychologie des Haushuhns. Z. Psychol. 88: 225-252.

Schleidt, W. 1973. Tonic communication: Continual effects of discrete signs in animal communication systems. J. Theoret. Biol. 42: 359-386.

_____. 1974. How "fixed" is the fixed action pattern? Z. Tierpsychol. 36: 184-211.

_____ and M.D. Shalter 1973. Stereotypy of a fixed action pattern during ontogeny in Coturnix coturnix coturnix. Z. Tierpsychol. 33: 35-37.

Schmidt, R.S. 1955. Termite (Apicotermes) nests: Important ethological material. Behaviour 8: 344-356.

Scott, J.P. 1967. The evolution of social behavior in dogs and wolves. Amer. Zool. 7: 373-381.

_____ and J.L. Fuller. 1965. Genetics and the Social Behavior of the Dog. Univ. of Chicago Press, Chicago.

Sibly, R.M. and D.J. McFarland. 1976. On the fitness of

behaviour sequences. Amer. Nat. 110: 601-617.

Sibley, C.G. 1957. The evolutionary and taxonomic significance of sexual dimorphism and hybridization in birds.
Condor 59: 166-191.

Siegel, S. 1956. Nonparametric Statistics for the Behavioral
Sciences. McGraw-Hill Book Co., New York.

Silver, H. and W.T. Silver. 1969. Growth and behavior of the
coyote-like canid of northern New England with observations on canid hybrids. Wildl. Mono. 17: 1-41.

Simmons, K.E.L. 1957. The taxonomic significance of head-
scratching methods of birds. Ibis 9: 178-181.

Simpson, M.J.A. 1969. The display of the siamese fighting
fish, Betta splendens. Anim. Behav. Mono. 1: 1-73.

_____. 1973. Social displays and the recognition of indiv-
iduals. p. 225-279 in P.P.G. Bateson and P. Klopfer
(eds.), Perspectives in Ethology, Vol. 1, Plenum Press,
New York.

Slater, P.J.B. 1973. Describing sequences of behavior.
p. 131-153, ibid.

_____. 1975. Temporal patterning and the causation of bird
behaviour. p. 11-33 in P. Wright, P.G. Caryl, and D.M.
Vowles (eds.) Neural and Endocrine Aspects of Behavior in
Birds, Elsevier Sci. Pub. Co., Amsterdam.

_____ and J.C. Ollason. 1972. The temporal pattern of behaviour in isolated male zebra finches: Transition analysis.
Behaviour 42: 248-269.

Sneath, P.H.A., and R.R. Sokal. 1973. Numerical Taxonomy:
The Principles and Practice of Numerical Classification.
W.H. Freeman & Co., San Francisco.

de Solla Price, D.J. 1960. Review of I. Asimov's the
intelligent man's guide to science. Science 132: 1830-
1831.

Sohn, J.J. 1977. Socially induced inhibition of genetically
determined maturation in the platyfish, Xiphophorus
maculatus. Science 195: 199-201.

Spieth, H.T. 1952. Mating behavior within the genus
Drosophila (Diptera). Bull. Amer. Mus. Nat. Hist. 99:
399-474.

Spigel, I.M, and D. Fraser. 1974. "Dominance" in the laboratory rat: The emergence of grooming. Z. Tierpsychol. 34:
59-69.

Stamps, J.A. and G.W. Barlow. 1973. Variation and stereotypy
in the displays of Anolis aeneus (Sauria: Iguanidae).
Behaviour 47: 67-94.

Stevenson, M.F. and T.B. Poole. 1976. An ethogram of the
common marmoset (Calithrix jacchus jacchus): General
behavioural repertoire. Anim. Behav. 24: 428-451.

Still, A.W. 1976. An evaluation of the use of Markov models
to describe the behaviour of rats at a choice point.

Anim. Behav. 24: 498-506.
Stokes, A.W. 1962. Agonistic behaviour among blue tits at a
    winter feeding station. Behaviour 19: 118-138.
Stokes, B. 1955. Behaviour as a means of identifying two
    closely-allied species of gall midges. Brit. J. Anim.
    Behav. 3: 154-157.
Struhsaker, T.T. 1970. Phylogenetic implications of some
    vocalizations of Cercopithecus monkeys. p.365-444 in J.R.
    Napier and P.H. Napier (eds.) Old World Monkeys, Academic
    Press, New York.
Suppes, P. and R.C. Atkinson. 1960. Markov Learning Models
    for Multiperson Interactions. Stanford Univ. Press,
    Stanford, Calif.
Syme, G.J. 1973. Competitive orders as measures of social
    dominance. Anim. Behav. 22: 931-940.
van Tets, G.G. 1965. A comparative study of some social
    communication patterns in the Pelicaniformes. Ornithol.
    Mono. 2: 1-88.
Tinbergen, N. 1959. Comparative studies of the behaviour of
    gulls (Laridae): A progress report. Behaviour 15: 1-70.
_____. 1960. The evolution of behavior in gulls. Sci. Amer.
    203: 118-130.
Verbenne, G. and P Leyhausen. 1976. Marking behaviour of some
    Viverridae and Felidae: Time-interval analysis of the
    marking pattern. Behaviour 58: 192-256.
Vierke, J. 1975. Beiträge zur ethologie und phylogenie der
    Familie Belontiidae (Anabantoidei, Pisces). Z.
    Tierpsychol. 38: 163-199.
Watson, A. and R. Moss. 1970. Dominance, spacing behaviour
    and aggression in relation to population limitation in
    vertebrates. p. 167-220 in A. Watson (ed.) Animal Popula-
    tions in Relation to their Food Resources, Blackwell Sci.
    Publ., Oxford and Edinburgh.
Westby, G.W. and H.O. Box. 1970. Prediction of dominance in
    social groups of the electric fish, Gymnotus carapo.
    Psychon. Sci. 21: 181-183.
Whitman, C.O. 1919. The behavior of pigeons. Pub. Carnegie
    Inst., Washington 257: 1-161.
Wickler, W. 1976. The ethological analysis of attachment.
    Z. Tierpsychol. 42: 12-28.
Wiley, R.H. 1973. The strut display of male sage grouse: A
    "fixed" action pattern. Behaviour 47: 129-152.
_____. 1975. Multidimensional variation in an avian display:
    Implications for social communication. Science 190:
    482-483.
_____. 1976. Communication and spatial relationships in a
    colony of common grackles. Anim. Behav. 24: 570-584.
Wilkinson, R. and P.E. Howse. 1975. Variation in the tempor-
    al characteristics of the vocalizations of bullfinches,

Pyrrhula pyrrhula. Z. Tierpsychol. 38: 200-211.

Willows, A.O.D., D.A. Dorsett, and G. Hoyle. 1973a,b. The neuronal basis of behavior in Tritonia. I. Functional organization of the central nervous system. III. Neuronal mechanism of a fixed action pattern. J. Neurobiol. 4: 207-237 & 255-285.

_____ and G. Hoyle. 1969. Neuronal network triggering a fixed action pattern. Science 166: 1549-1551.

Wilson, E.O. 1975. Sociobiology: The New Synthesis. Harvard Univ. Press, Cambridge, Mass.

Wilson, W.L. and C.C. Wilson. 1975. Species-specific vocalizations and the determination of phylogenetic affinities of the Presbytis aygula-melaophos group in Sumatra. 5th Internat'l. Congr. Primat., 459-463.

Wolfe, J.L. 1966. Agonistic behavior and domestic relationships of the Eastern chipmunk, Tamias striatus. Amer. Midl. Nat. 76: 190-200.

INFORMATION THEORY AS AN ETHOLOGICAL TOOL

June B. Steinberg

University of Illinois at Chicago Circle

Abstract: The use of information theory enables the student of animal communication to investigate such important questions as: What are the probablistic constraints placed on the behavior of a recipient animal by the acts of a signaler? How important are an animal's own preceding acts in determining his subsequent behavior? How much information is carried by specific acts? How do signals differ in effect as the result of such factors as size of signaler and context of the signal and how do these factors interact?

To make proper use of this tool one must become familiar with the most commonly used information theoretic measures. These are H(B), a measure of information present; H(B/A), a measure of the information present in animal B's acts when A's acts are known; T(A:B), the normalized transmission; and h_t, the contribution of a particular act to the overall transmission. Recent advances in this field have provided ways to decrease the bias of estimates of information measures and to estimate their variance. These advances now permit statistical comparison of these measures.

INTRODUCTION

In 1965 two landmark papers were published which used information theory to analyze animal behavior. One dealt with the sequential constraints in social communication of rhesus monkeys (Altmann, 1965) and the other with the analysis of aggressive encounters between hermit crabs (Hazlett and Bossert, 1965). Despite certain limitations, these papers helped pave the way for others to use this valuable tool and in recent years there have been several studies in which information theoretic measures have been used to analyze sequences of behavior (e.g., Dingle, 1969, 1972; Chatfield and Lemon, 1970; Hazlett and Estabrook, 1974a, 1974b; Rubenstein and Hazlett, 1974; Steinberg and Conant, 1974; Hyatt, In Prep.).

It is important for anyone interested in quantitative methods of investigating animal behavior to be aware of the biological meaning of information theoretic measures and the advantages to be gained by their use. These methods of analysis are not without drawbacks, but when one is aware of the limitations, the formulations of information theory offer a powerful aid in sequential analysis and the understanding of the constraints placed on the behavior of one animal by another when communication occurs between them.

Information theory is valuable because it permits the study of both inter- and intra-individual communication. We can test the hypothesis that communication occurs between animals (inter-individual communication) and quantify the amount of that communication. When an animal makes specific sounds or movements in the presence of conspecifics it is tempting to assume that it is communicating. Unless, however, these acts can be shown to change the probability distribution of acts by conspecific observers of such behavior, no such assumption is warranted. Too often communication is subjectively assumed instead of being objectively tested. Information theory, along with other methods, makes the latter course of action possible. Furthermore, we can estimate the amount by which an individual's next act is the result of its own immediately preceding act or acts (intra-individual communication). Dingle (1969) investigated such sequences and found that intra-individual and inter-individual communication were equally important in mantis shrimp. This result warns us against ignoring the part played by sequential constraints on the individual's own behavior. Another feature of information theory is that with appropriate caution, comparisons of the levels or rates of communication in different species or situations can be made.

With modern portable equipment, such as video-tape recorders and electronic data recorders, ethologists can now, in many cases, collect the data necessary for informational analysis directly from observations in the natural habitat. The formulations of information theory make it possible to compare such field-collected data with laboratory observations and also to test statistically whether confined conditions sufficiently mimic the "real world" of the animals to justify the use of data collected under such conditions.

Not only is it possible to quantify the amount of information an animal transmits to an observer, it is also possible to rank the signals used by that animal in terms of their contribution to that transmission (Steinberg and Conant, 1974). Such a list is valuable because it pinpoints those signals carrying the most information and makes it possible to examine these signals more closely. From such data a much clearer picture of the total communicative repertoire of the animal emerges.

The information theoretic model can be extended to quantify inherent variability in both static (e.g., size, sex, weight, and color) and behavioral aspects of interactions (e.g., movements, signals, and sounds) without the necessity of separating these various factors artificially and analyzing one while holding the other constant. This is done by the technique of character analysis (Estabrook, 1967) and has proved very effective in analyzing the agonistic behavior of spider crabs (Hazlett and Estabrook, 1974a), hermit crabs (Hazlett and Estabrook, 1974b), and crayfish (Rubenstein and Hazlett, 1974).

One final advantage of this method of analysis should be pointed out before proceeding to a discussion of the formulations themselves. The data gathering process needed to perform an informational analysis is a long one. Particularly if one does such a study on field-gathered data it will be necessary to spend many long and often tedious hours in data collection in order to gather sufficient data for the analysis. During this time several corollary and generally unsuspected benefits accrue to the investigator, who will be rewarded for the extra time spent by gaining a knowledge of the animal far exceeding that procured by more traditional methods. Furthermore, the hours spent observing and waiting for something to happen will usually lead to insights and testable hypotheses which might never have been gained had not so many interactive sequences been recorded.

I. DYADIC INTERACTIONS

A. Information

As formulated by Shannon and Weaver (1949), the unit of information, or _bit_, is equivalent to the amount of information required to choose between two equiprobable alternatives. Where there are four equally probable alternatives a choice would require two bits of information and with n alternatives $\log_2 n$ bits. It follows that in a simple system, with all outcomes equally probable, determination of one event requires $\log_2 n$ bits of information and if one is told which event actually occurred, that is the amount of information received. However, animals do not choose acts from their behavioral repertoire with equal probability, therefore the probability of occurrence of an act must be considered when determining the amount of information. The formula for calculating information (H) is

$$H(X) = - \sum_{i=1}^{n} P(i)\log_2 P(i)$$

where X is made up of $\underline{n}$ categories $(x_1, x_2, \ldots, x_n)$, and $p(i)$ is the probability of the ith value of X. In practice the actual probabilities are rarely known and are calculated from the maximum likelihood estimator:

$$\hat{P}(i) = \frac{\text{no. of occurrences of event } i}{\text{total no. of occurrences of any event}}.$$

To simplify computation, the following formula is useful:

$$\hat{H}(x) = \frac{1}{N} [N \log_2 N - \sum_{i=1}^{n} n_i \log_2 n_i]$$

where N is the total number of occurrences of all events and n_i the number of occurrences of event i.

In some circumstances it may be convenient to consider the variable X as a character (e.g., sex, color or behavioral repertoire) with $\underline{n}$ states $(x_1, x_2, \ldots x_n)$ (Estabrook, 1967; In Press). Throughout this paper characters (variables) will be referred to by capital letters and the states of a character (events) by lower case letters. The states of a character correspond to the events mentioned above and a character may have only two states (as with "sex") or many (as with "behavioral repertoire"). Characters, such as sex, can be considered in several ways. For example, one might be concerned with the sex of two interacting animals. This could be considered by a character with two states: x_1 = both animals the same sex; x_2 = sex of the animals differs; or the character could have three states:

x_1 = one animal male and the other female
x_2 = both animals male
x_3 = both animals female

Another consideration of this character might involve four states:

x_1 = first animal male, second animal female
x_2 = first animal female, second animal male
x_3 = both animals female
x_4 = both animals male

The choice of characters to investigate becomes virtually limitless; it is bounded only by the ability of the investigator to ask good questions. Selecting those characters for analysis that lead to fruitful insights into the behavior of an animal becomes the major task of this form of analysis and,

TABLE 1

Hypothetical Distribution of Behavioral
States (X) and Sex (Y). The Independent
Distributions of X and Y are Shown in (A) while
Distributions of Behavioral States by Sex are Shown in (B).

A. Frequencies of States of X	Frequencies of States of Y
$x_1 = 22$	$y_1 = 29$
$x_2 = 21$	$y_2 = 71$
$x_3 = 57$	

Total observations = 100 in which the character states x_1 and y_1 were recorded.

B. Distribution of Acts as Performed by Males or Females

Acts by Females (y_1)	Acts by Males (y_2)
$x_1 = 10$	$x_1 = 12$
$x_2 = 8$	$x_2 = 13$
$x_3 = 11$	$x_3 = 46$

if the characters are indeed prudently selected, the inter-relationship between static and dynamic (behavioral) characters will begin to be understood. For further discussion of character analysis see Estabrook (In Press).

An example of the way in which the information of characters can be estimated is presented in Table 1. Let X = the character "an animal's behavioral repertoire". For simplicity we will select a mythical species with only three acts in that repertoire. The character X can then assume the following states:

$$x_1 = act\ 1$$
$$x_2 = act\ 2$$
$$x_3 = act\ 3$$

In 100 observations the observed frequencies of the states of X are shown in Table 1A. The information associated with a particular set of numbers, discarding the name of the variable is often denoted as follows:

$\hat{H}(X) = H(22,21,57)$ then

$\hat{H}(X) = \frac{1}{100} (664.38 - (98.11 + 92.24 + 332.47)) = 1.146$ bits.

Now let us consider another character, Y, with two character states. The character Y could represent sex so that:

$$y_1 = \text{act by a female}$$
$$y_2 = \text{act by a male}$$

The frequencies of Y in the 100 observations also are shown in Table 1A and the information associated with Y is denoted as follows:

$$\hat{H}(Y) = H(29,71) = \frac{1}{100} (664.38 - (140.88 + 436.63)) = 0.869$$
$$\text{bits}$$

B. Multivariable Information

For the vector variable (X,Y) with representative values $x_i y_j$ which occurs with probability P_{ij} or occurrence n_{ij} the information is:

$$H(X,Y) = - \sum_{j=1}^{n} \sum_{i=1}^{n} P_{ij} \log_2 P_{ij}$$

$$\hat{H}(X,Y) = \frac{1}{n} [N \log_2 N - \sum_{j=1}^{n} \sum_{i=1}^{n} n_{ij} \log_2 n_{ij}]$$

i.e., exactly as for a single variable. This can easily be extended to $\underline{n}$ dimensions. Returning to the data set in Table 1, we can now consider the "joint" information distributed as in Table 1B. Therefore

$$\hat{H}(X,Y) = H(10,8,11,12,13,46) = 2.239 \text{ bits.}$$

It is often convenient to place such distributions in a contingency table with the states of one character as the rows and the states of another as the columns as in Table 2. In such cases the marginal distributions represent the distribution of each character and the cell entries represent the joint distribution. Such contingency tables can be extended to $\underline{n}$ dimensions whenever the situation warrants.

Often in the literature one finds contingency tables where the rows represent the first act of a dyad between two animals and the columns represent the second, or following act (Hazlett and Bossert, 1965; Dingle, 1969; Steinberg and Conant, 1975). Steinberg and Conant generated two tables from such dyads, in order to preserve the distinction between initiators and responders and found this distinction important, as did Hazlett and Estabrook (1974a).

Nevertheless there is a problem in dealing with a behavioral interaction of the sort $a_1 b_2 a_1 a_1 a_2 b_1 b_2 b_1 \ldots$, where the letters represent animals A and B and the numbers represent acts 1 and 2 from each animal's repertoire. Animals do not conveniently alternate acts so that they can be entered

TABLE 2

*Rearrangement of the Data of Table 1
into a Contingency Table*

		Y			
		y_1	y_2	$\Sigma=$	
X	x_1	10	12	22	x_1
	x_2	8	13	21	x_2
	x_3	11	46	57	x_3
	$\Sigma=$	29	71		
		y_1	y_2		

easily and unambiguously in a contingency table. There are several ways of dealing with such situations. One can look only for those instances where an act by A is followed by an act by B and ignore the rest. This gives two A-B dyads from the above string (a_1 b_2 and a_2 b_1) and one B-A dyad (b_2 a_1). Potentially important parts of the interaction, where one animal does several things in a row before there is a discernible change in the behavior of the other, are lost with this system. As a remedy, the A-A and B-B transition can be handled in separate tables, as if they are examples of intra-individual communication.

Such treatments ignore the possibility that the sequence a_1 a_1 a_2 from the above sample might not have been performed if A were alone. The string is probably the combined result of A's own tendency to act once it has just acted (intra-individual communication) and B's presence (inter-individual communication). A way to treat the inter-individual aspects of this situation is to insert a third act, "quiescence". The sample string would appear

$$a_1 \ b_2 \ a_1 \ b_3 \ a_1 \ b_3 \ a_2 \ b_1 \ a_3 \ b_2 \ a_3 \ b_1 \ \cdots$$

Now there is an alternation of actors and it is easy to construct contingency tables in the A-B and B-A directions. The danger of artificially inflating the value of "quiescence" in such situations is far less than the danger of distorting the data by ignoring much of the behavior, or treating it solely as if the other animal were absent. Intra-individual com-

TABLE 3

Representation of Hypothetical Sequence of
Behavioral States in Matrix Form (See Text)

		Y			
		a_1	a_2	b_1	b_2
X	a_1	1	1		1
	a_2				
	b_1				1
	b_2	1		1	

munication can still be dealt with by examining the entire string of A acts and B acts separately. This would give the sequence $a_1 \, a_1 \, a_1 \, a_2...$ for A, and $b_2 \, b_1 \, b_2 \, b_1...$ for B. The entries in the A-A table would be $a_1 \, a_1 \, a_1 \, a_2...$ and the B entries would be similarly determined.

Alternatively all of the transitions can be placed in a single table (Baylis, 1975). For the sample string such a table would look like Table 3. This matrix avoids the difficulties discussed above, but presents us with new ones. We no longer can measure the transmission between A and B, but are now concerned with 2 different variables or characters. The X character is "actor and first act in a dyad" and the Y character is "actor and second act in a dyad". Furthermore when one is dealing with an animal with a repertoire of 10 acts (a rather small number) the matrix will be 20 x 20 or four times as large as matrices constructed by other methods and often require a large number of dyads to provide statistical validity to any analysis (see discussion on sample size, below).

Interactive sequences where there is an alternation of actors have been analyzed by examining information as the interactions proceed (Steinberg and Conant, 1974; Hazlett and Estabrook, 1974a, 1974b; Rubenstein and Hazlett, 1974). In such cases all first acts of interactive sequences are treated as a character with the number of states equal to the number of acts in the repertoire. The second acts of interactions are another character, etc. Hazlett and his colleagues have also started at the end of interactions and worked backwards. This avoids the problem of dealing with situations where a character (e.g., act 8) is composed of the final acts of some

interactions and the middle acts of others, since all interactions are not of the same length. A problem is that the
most interesting part of the sequence is likely to be in the
middle. This is where uncertainty is greatest and where con-
flicts are probably settled. Much of this is missed by looking only at the ends of sequences. An example of the insights
gained from such analyses can be taken from Rubenstein and
Hazlett (1974). They found that in long fights between crayfish the sixth act (third act by defender) reduced uncertainty
about the eventual winner of a fight by 44%. There are many
other such conclusions in these papers and the interested
reader is referred to them.

The maximum value which can be taken by H(X) is reached
when all of the states of X are equiprobable. In such cases
H_{max} is simply equal to the base 2 log of the number of
states, expressed in bits. This number represents the average
number of yes/no questions needed to determine the state of X.
H_{max} is clearly dependent on our choice of the categories into
which a character is divided. The number of states in an
animal's behavioral repertoire may be as much a function of
the observer determining that repertoire as it is of the
animals themselves. It is therefore necessary to be extremely
careful in determining the states of a character and wary when
absolute values of information measures are compared (Stein-
bert and Conant, 1974). Identifying units of behavior is an
extremely important task confronting all behavioral biologists. The interested reader is referred to Altmann (1967),
Conant and Steinberg (1973) and Manning (1976) for various
approaches to this problem. The minimum value of H(X) is 0;
it can never be negative. H(X) will be 0 when X always
assumes one of its states, e.g., x_1, and never assumes any of
the others. Zero yes/no questions are necessary to determine
the state of X, since it is always known. In such a case we
receive no information when we learn that state x_1 has
occurred, since we knew in advance that it would.

H is a measure of randomness or variability in a distribution. If we consider X, an animal's behavioral repertoire,
a high value of H(X) means that there is much apparent randomness and little predictability in the way in which the animals
are choosing behavior from this repertoire. Our uncertainty
about what an animal will do is very high and we receive much
information when we learn what it has done. A low value H(X)
means just the opposite.

The statistically minded reader may find it convenient
to think of H much as he or she thinks of variance. The
equations are different, but since anything that increases
variance will also increase the amount of information, the
conceptualization is appropriate. The advantage of dealing
with information rather than variance accrues from the fact

that the information in a discrete statistical distribution
is independent of the unit of measurement. We can therefore
deal with distributions where there is no metric and we would
not ordinarily be able to use variance. Furthermore H can be
used for making comparisons in situations where variances cannot be compared because they are based on different metrics.

C. Conditional Information

To calculate information conditional on a particular
value or state of a character, such as $H(X/y_2)$ (which reads
"the information of the character X given that character Y
assumes state 2"), simply use the definition for H with that
portion of the distribution associated with the particular
state. Referring again to our sample data in Tables 1B and 2
we have:

$$\hat{H}(X/y_2) = H(12,13,46) = 1.288 \text{ bits}$$
$$\hat{H}(X/y_1) = H(10,8,11) = 1.573 \text{ bits}$$

What is usually meant by conditional uncertainty, however, is a weighted average of the information conditional on
particular values. The conditional information is weighted by
the probability of a given event or character state. Thus:

$$\hat{H}(X/Y) = \frac{29}{100} H(X/y_1) + \frac{71}{100} H(Y/x_2) = \frac{29}{100} (1.573) +$$

$$\frac{71}{100} (1.288) = 1.371 \text{ bits}$$

$$\hat{H}(Y/X) = \frac{22}{100} H(10,12) + \frac{21}{100} H(8,13) + \frac{51}{100} H(11,46) =$$

$$0.219 + 0.201 + 0.403 = 0.823 \text{ bits}$$

H(Y/X) can also be calculated this way:

$$H(Y/X) = H(X,Y) - H(X)$$

The conditional information, H(Y/X), measures the average information remaining unknown about Y when X has been determined. It can never be greater than H(Y), which means that
on the average we cannot have less information about Y when
we know X than we had before we were provided with such knowledge. When knowledge of X provides no information about Y,
H(Y/X) will be equal to H(Y). On the average, when knowledge
of X increases the predictability of Y, H(Y/X) will be smaller
than H(Y); and when Y is completely determined by knowledge of

X, H(Y/X) will be 0.

D. Transmission

    The difference between H(Y) and H(Y/X) is the trans-
mission, T(X;Y). This is the measure referred to by Estabrook
(1967; In Press) and Hazlett and his coworkers (Hazlett and
Estabrook, 1974a, 1974b; Rubenstein and Hazlett, 1974) as
redundancy. The transmission, which is never negative and
never exceeds H(Y) or H(X) is a measure of relatedness or
shared information. In a situation involving animal communication it will equal 0 if, and only if, the immediate subsequent actions of the two animals are statistically independent, i.e., they share no information in common. In such a
case, knowledge of what one animal does tells nothing about
what another animal will do (the probability distribution of
acts by the second animal remains unchanged by the acts of the
first). We are usually justified in assuming that no communication has occurred in such instances. It is, of course,
possible for an animal to receive a message, store it away,
and act on it at a later time or even choose to ignore it. In
such cases there will be no observable change in behavior on
the part of the recipient animal, but communication between
the two will have occurred. Our analyses will be inaccurate
to the extent that such delayed reactions occur and we should
always be aware of this problem and searching for evidence of
it. If a data set is large enough, it is possible to look for
higher order transmission, thus determine at least short term
"storage" of received transmission. Until, however, it is
possible to monitor the nervous systems of recipient animals
and gain direct evidence for received information, the best
way to tell if communication exists is to find a change in the
probability distribution of acts by one animal as the apparent
result of acts by another. The amount by which T(X;Y) exceeds
0 and approaches H(Y) is a measure of the constraints between
the variables or characters X and Y. In the case of communicating animals it is an indication that the acts of one
animal are having considerable effects upon the subsequent
behavior of another, there is much shared information in the
distribution of acts and communication is probably occurring.
    T(X;Y) is calculated by the following formulas:

$$T(X;Y) = H(Y) - H(Y/X) = H(X) - H(X/Y) = H(X) + H(Y) - H(X,Y)$$

Returning again to our example in Tab. 1B we have

$$\hat{T}(X;Y) = 1.416 + 0.869 - 2.239 = 0.046 \text{ bits}$$

E. Normalized Transmission (% uncertainty reduction)

The normalized transmission is calculated from the ratio

$$t(X;Y) = \frac{T(X;Y)}{H(Y)}$$

When expressed as a percentage t represents the percent by which uncertainty about Y is reduced by knowing X. In our example t(X;Y) = 3.2% while t(Y;X) = 5.3%. These rather low values tell us that very little information is shared by characters X and Y. In other words, knowing the sex of an animal is, in this instance, a poor predictor of what an animal will do. Knowing what an animal has done is only slightly (2%) better in providing information as to the sex of that animal.

When calculating t for data involving interactions between male grasshoppers, Steinberg and Conant (1974) found that 22% of the uncertainty about a responding animal was reduced by knowing the act of the initiating animal, but there was only a 12% reduction in the reverse direction (i.e., when responders communicated with initiators). In this case initiators are affecting responders more than responders are affecting initiators. Rubenstein and Hazlett (1974) showed that "the positions adopted by a crayfish seem to influence its own subsequent behavior" (p. 206); also see Oden, this volume. This effect was found to be less for initiators than respondents. In the former case there was 26% uncertainty reduction, while in the latter case there was a 50% reduction. These results are mentioned to show the intuitive power of this measure. One need not be at all mathematically sophisticated to understand that a 50% reduction in uncertainty is a great deal of information. Two characters which reduce uncertainty about each other by that much are obviously tightly linked and the one constrains the other considerably.

It should be obvious that t can range from 0 to 1 (there can be complete independence of the two characters or complete linking of them). Since this relative measure is largely independent of the number of categories (character states) used in the calculation of the previously discussed absolute information measures, it is the measure most recommended for comparisons. Characters which reduce uncertainty by 50% are, in an informational sense, quite comparable. This is true even if the situation involves quite different sets of characters. In one case we might be considering the linking of "being a respondent crayfish" to inter-individual communication, and in the other we might be considering the next-to-last acts in interactive sequences of male and female grasshoppers. Since the same amount of uncertainty is reduced by each pair of characters, they are similar in the amount of information they share with each other even though the H values might be quite

different in the two cases. This ability to compare dis-
similar measures is an important feature of information
theory.

F.   Amount of Information Per Character State

     Blachman (1968) defined a measure which gives the rela-
tive contribution of each character state to the overall information, H.   The formula for this measure is:

$$J(x;Y) = \sum_{y} P(x/y) \ \log_2 \frac{P(y/x)}{P(y)}$$

A difficulty with this measure is that it may be large for an
event "x" that is very rare.   In such cases the validity of
the measure is questionable.   To avoid this problem we can
calculate the weighted average of the information carried by
each event.   The result of this calculation is the contribution of each event to the overall transmission between X and
Y.   The formula for this is

$$T(X;Y) = \sum_{x} p(x)J(x;Y)$$

Steinberg and Conant (1974) ranked the signals of grasshoppers
in terms of their contribution to the transmission between
males and the reader is referred to that paper for a discussion of the specific insights provided by such a ranking.
     To see how $J(x;Y)$ is computed, we shall return to our
hypothetical animal with its three-act repertoire.   For the
sake of variety we shall assume that we have a different set
of observations in which the behavior of interacting males was
recorded.   We have 100 dyads in which an act by one male was
followed by an act of another.   These data are displayed in
Table 4 in the form of a contingency table where the rows are
the first acts and the columns are the following acts of these
dyads.
     The information measures which have already been dis-
cussed can be estimated as follows:

$$\hat{H}(X) = H(40,20,40) = 1.5219 \text{ bits}$$
$$\hat{H}(Y) = H(30,60,10) = 1.2955 \text{ bits}$$
$$\hat{H}(X,Y) = H(10,30,20,20,10,10) = 2.4464 \text{ bits}$$
$$\hat{H}(Y/X) = \hat{H}(X,Y) - \hat{H}(X) = 0.9245 \text{ bits}$$
$$\hat{T}(X;Y) = \hat{H}(X) + \hat{H}(Y) - \hat{H}(X,Y) = 0.371 \text{ bits}$$

The probabilities needed to calculate $J(x;Y)$ are estimated in
the following ways:

TABLE 4

Contingency Table of Hypothetical
Dyads of Behavioral Acts

	y_1	y_2	y_3	$\Sigma=$
x_1	10	30	0	40
x_2	0	20	0	20
x_3	20	10	10	40
$\Sigma=30$		60	10	100

$P(y/x) = \dfrac{P(x,y)}{P(x)}$ $P(x,y) = P$ (joint occurrence of x & y)

$\hat{P}(x)$ is estimated by $\dfrac{\text{row total}}{N}$

$\hat{P}(y/x)$ is estimated by $\dfrac{\text{cell entry}/N}{\text{row entry}/N} = \dfrac{\text{cell entry}}{\text{row total}}$

$\hat{P}(y)$ is estimated by $\dfrac{\text{column total}}{N}$

Ranking x's signals in terms of their relative contributions to the overall transmission results in a reverse of the original order: i.e., x_3 x_2 x_1. Note that ranking the signals according to their contributions to H(Y) results in the order: x_2 x_3 x_1. This arrangement overvalues the rare act, x_2, and is not as valid a measure as $p(x)J(x;Y)$ since the rare act is not contributing as much to the transmission as are other acts. The biological importance of rare acts is, of course, another matter.

II. MULTI-INDIVIDUAL INTERACTIONS

It should be clear by now that there is no great difficulty in calculating information measures for situations where two animals interact or where one is concerned about the relationship between two characters. But what do we do about those situations in which three or more animals interact? If our concern is the level of stochastic dependence of sequences of behavior generated under such circumstances and we do not particularly care about which animal performed a given act in the string we could use Altmann's (1965) method. He analyzed sequential data on rhesus monkeys so that he could estimate the amount of information which knowing one act in the string

TABLE 5

*Calculation of Estimated Probabilities
from Data Set in Table 4*

$P(y/x)$	$\dfrac{P(y/x)}{P(y)}$	$\log_2 \dfrac{P(y/x)}{P(y)}$	$J(x;Y)$	$P(x)J(x;Y)$
$P(y_1/x_1)=0.25$	0.833	-0.263	-0.066	
$P(y_2/x_1)=0.75$	1.250	0.322	0.241	
$P(Y_3/x_1)=0$	0	0	0.0	
			$\overline{0.175}=J(x_1;Y)$	0.070
$P(y_1/x_2)=0$	0	0	0	
$P(y_2/x_2)=1$	1.667	0.737	0.737	
$P(y_3/x_2)=0$	0	0	0	
			$\overline{0.737}=J(x_2;Y)$	0.147
$P(y_1/x_3)=0.5$	1.667	0.737	0.369	
$P(y_2/x_3)=0.25$	0.417	-1.263	-0.316	
$P(y_3/x_3)=0.25$	2.500	1.322	0.331	
			$\overline{0.384}=J(x_3;Y)$	0.154

$\Sigma\ J(x;Y) = 0.175 + 0.737 + 0.384 = 1.296$ bits

$\Sigma\ P(x)J(x;Y) = 0.371$ bits $= T(x;Y)$

gave about what the next act would be. He then went back a
step and estimated the information gained by knowing the preceding two acts in the string, continuing this procedure for
three and four predecing acts.

A problem with this type of analysis, is that as the number of preceding acts one is considering increases, the data
base for estimating probabilities decreases. In Altmann's
study (1965) there were 120 acts in the repertoire of the monkeys and he averaged 45.9 observations of each act. When
four-act quadrads were considered, there were 120^4 possible
quadrads and he obtained only 13.8×10^{-6} observations per
quadrad. The logical outcomes of this difficulty are discussed by Altmann, who stated (pp. 515-516), "In theory, this
procedure of looking farther back into the history of the
interaction process is continued until no further reduction in
the uncertainty of our predictions of behaviour is obtained.
In practice, however, it is extremely difficult to get an
adequate sample for an animal that has such a large repertoire. Yet this is enigmatic, for the monkeys are faced with
just the same problem: to whatever extent their choice of a

course of action in a particular situation is based upon
their past experiences with similar situations, they are
limited to the sample that can be obtained within their life-
time, particularly within the first few years... For an animal with a repertoire as large as that of rhesus monkeys, the
number of possible courses of interaction is so large for even
moderately short sequences that it is very hard to believe
that the monkeys utilize very many of them."

In their paper on interactions of male grasshoppers,
Steinberg and Conant (1974) showed that initiators and responders were transmitting different amounts of information even
though they shared a common and limited behavioral repertoire.
It is easy to imagine a situation where there are three interactors, e.g., a large male, a small male and a female each
employing the same signals, but sending different information.
To analyze such a situation one would have to observe and record many sequences of such interactions (we will deal with
the problem of adequate sample size below).

A small segment from a typical string of interactions
among our mythical creatures, with their three-act repertoires, might look like this:

$$a_1\ b_2\ c_1\ b_2\ b_1\ a_2\ b_1\ c_2\ a_2\ b_1\ c_2\ c_1\ b_2...$$

where the letters represent interactors and the numbers represent acts. In this case let us assume that A is a large
male, B is a small male and C is a female. Only two acts
occur in the string because the third act in this animal's
repertoire is "quiescence" (see below). There are six
different three-dimensional contingency tables which can be
generated from this string if we are concerned with maintaining each participant separately. We could look for an act by
A followed by an act by B followed by an act by C (ABC), or
act by A followed by C followed by B (ACB) or BAC, etc. To
generate a list of ABC triads from the sample string we proceed down the string as follows: The first three acts are
already in the ABC order so they are listed as they occurred,
$a_1\ b_2\ c_1$. The fourth event in the string is "b_2". To
preserve the order of "act by A, act by B, act by C" we can
now insert our third act "quiescence" and list this triad as
$a_3\ b_2\ c_3$, remembering that when we do this we assume that B
might not have performed act 2 at that time if A and C were
not there. Since they did not "do" anything that we could
observe, we record them as "quiescent". Proceeding down the
string we generate six triads in the ABC direction. To generate the triads in the ACB direction, notice that the sequence begins with $a_1\ b_2$. There is no intermediate act by C
so c_3 is inserted, giving $a_1\ c_3\ b_2$ as the first ACB triad.
The next two acts in the string are $c_1\ b_2$ and it is necessary

TABLE 6

*Various Triads of Behavioral Acts, Obtained
from a Sequence of Acts by Three Animals,
Dependent Upon the Sequence of Actors*

$a_1\ b_2\ c_1\ b_2\ b_1\ a_2\ b_1\ c_2\ a_2\ b_1\ c_2\ c_1\ b_2\ldots$

ABC	ACB	BAC	BCA	CAB	CBA
$a_1b_2c_1$	$a_1c_3b_2$	$b_3a_1c_3$	$b_3c_3a_1$	$c_3a_1b_2$	$c_3b_3a_1$
$a_3b_2c_3$	$a_3c_1b_2$	$b_2a_3c_1$	$b_2c_1a_3$	$c_1a_3b_2$	$c_3b_2a_3$
$a_3b_1c_3$	$a_3c_3b_1$	$b_2a_3c_3$	$b_2c_3a_3$	$c_3a_3b_1$	$c_1b_2a_3$
$a_2b_1c_2$	$a_2c_3b_1$	$b_1a_2c_3$	$b_1c_3a_2$	$c_3a_2b_1$	$c_3b_1a_2$
$a_3b_3c_2$	$a_3c_2b_3$	$b_1a_3c_2$	$b_1c_2a_2$	$c_2a_2b_1$	$c_3b_1a_3$
$a_3b_3c_1$	$a_3c_2b_3$	$b_3a_2c_3$	$b_1c_2a_3$	$c_2a_3b_3$	$c_2b_3a_2$
	$a_2c_3b_1$	$b_1a_3c_2$	$b_3c_1a_3$	$c_1a_3b_2$	$c_3b_1a_3$
	$a_3c_2b_3$	$b_3a_3c_2$	$b_2c_3a_3$		$c_2b_3a_3$
	$a_3c_1b_2$	$b_3a_3c_1$			
		$b_2a_3c_3$			

to insert a_3 to preserve the order. This method gives 9
triads in the ABC direction. Triads for the other orders of
interactors are generated in the same way and all of these are
listed in Table 6.

The six triads for the ABC direction can be placed into a
three-dimensional contingency table for analysis. Obviously
this small string is an insufficient sample to analyze by itself, but I have done so in Tab. 7 simply for demonstration.
In this matrix the first triad, '$a_1\ b_2\ c_1$' is placed in the
cell with the (1) in parentheses. The second triad,
$a_3\ b_2\ c_3$ is in cell (2) and it is hoped that the reader can
see without too much difficulty how the other triads are entered. Tables I-VI contain the same data in two- and one-
dimensional representations. Table I contains only the AB
dyads, ignoring C. Table II shows the original matrix
collapsed to one dimension and containing only the frequency
of A's acts. Similarly, Table III has AC, Table IV has C,

TABLE 7

The three tables in the top row comprise a three-dimensional contingency matrix where there are three variables: A = acts by large males, B = acts by small males, and C = acts by females. We can then place in this matrix the frequencies of triads where first A acted, followed by an act by B, followed by acts of C. Frequencies of the three possible acts by A (a_1 a_2 a_3) are entered in the rows of the matrix, acts by B (b_1 b_2 b_3) are entered in the columns of the matrix and acts by C (c_1 c_2 c_3) are recorded in the three tables of the matrix. See text for further explanation.

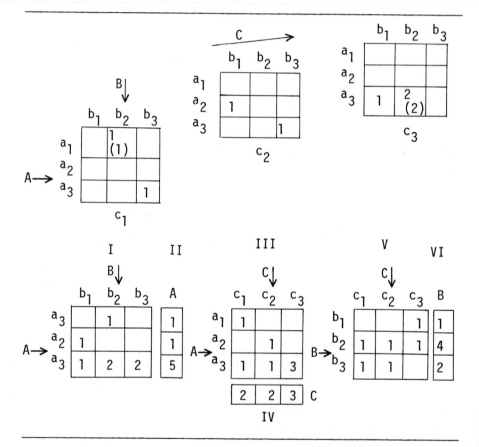

Table V has BC and Table VI has B frequencies.

The reader may want to calculate some of the information measure we have already described using the preceding matrix. In Table 8 are calculated H's, transmissions, and normalized transmissions (expressed as % uncertainty reduction) for all

TABLE 8

*Information Measures Calculated from the Data
Set Presented in Table 6*

	ABC	ACB	BAC	BCA	CAB	CBA
H(ABC)	2.52	2.20	2.92	2.75	2.52	2.73
H(A)	1.49	1.22	0.72	1.30	1.38	1.22
H(B)	1.56	1.59	1.57	1.44	1.45	1.59
H(C)	1.56	1.53	1.49	1.50	1.56	1.66
H(AB)	2.24	2.20	2.17	2.16	2.24	2.42
H(AC)	2.13	2.20	1.97	2.50	2.52	2.42
H(BC)	2.52	1.89	2.72	2.50	2.24	2.20
T(A;B;C)	2.09	2.14	0.86	1.49	1.87	1.74
T(A;B)	0.81	0.61	0.12	0.58	0.59	0.39
T(A;C)	0.92	0.55	0.24	0.30	0.42	0.46
T(B;C)	0.60	1.23	0.34	0.44	0.77	1.05
T(AB;C)	1.28	1.53	0.74	0.91	1.28	1.35
T(AC;B)	1.17	1.59	0.62	1.19	1.45	1.28
T(BC;A)	1.49	0.91	0.52	1.05	1.10	0.69
t(A;B)	52%	38%	8%	40%	41%	25%
t(B;A)	54%	50%	17%	45%	43%	32%
t(A;C)	59%	36%	16%	20%	27%	28%
t(C;A)	62%	45%	33%	23%	30%	38%
t(C;B)	38%	77%	22%	31%	53%	66%
t(B;C)	39%	80%	22%	29%	49%	63%
t(BC;A)	100%	75%	72%	73%	80%	57%
t(AB;C)	82%	100%	50%	61%	82%	81%
t(AC;B)	75%	100%	39%	83%	100%	81%

of the matrices generated from the lists of triads in Table 6.
These quantities can be linked by the following identities
which will facilitate computation.

$$T(A;B;C) = T(A;B) + T(AB;C) = T(A;C) + T(AC;B) =$$

$$T(B;C) + T(BC;A)$$

Now three of the transmissions are easily found by subtraction.
Not only does this table supply answers to check computations,
but several interesting observations are derived by examining
these calculations.

First of all it is apparent that the order in which we
consider the interactors makes a difference. The values are

different for each order. This is not surprising since the
lists were different in each case, but it merits mention.
Secondly, looking at the first column in this table, we see
that 100% of our uncertainty about A is removed by knowing
BC. To understand this, look at the information values in
this column and note that H(BC) = H(ABC). In other words,
there was no more information in knowing ABC than there was in
knowing BC; knowing BC completely determined A. Referring
back to the matrix for ABC triads, we see that if we know that
$b_1 c_3$ happened, the only possible "a" is a_3. When $b_2 c_1$
occurred, only a_1 is possible, etc.

Another matrix for handling the same interaction appears
in Table 9. This representation is of the form used by Baylis (1975). There are two variables (characters). The first
character is "first and second actors in a triad and their
acts", denoted by X, and the second character is "third actor
in a triad and its acts", denoted by Y. One value of such a
table might be to rank the acts and actors in terms of their
contributions to transmission. For example, it might be of
value to know that animal B doing act 2, followed by A doing
act 2, contributes maximally to the transmission to the third
animal in these triads.

The information measures for this matrix are presented
below. Again I caution that there is no pretense that this is
a sufficient sample size for estimating such measures; it is
done only as a heuristic illustration.

$$\hat{H}(X,Y) = H(1,2,1,1,1,1,1,1,1,1,1) = 3.42 \text{ bits}$$

$$\hat{H}(X) = H(1,2,1,2,1,1,1,1,1,1) = 3.25 \text{ bits}$$

$$\hat{H}(Y) = H(2,3,2,2,3) = 2.29 \text{ bits}$$

$$\hat{T}(X;Y) = 2.12 \text{ bits}$$

$$t(X;Y) = 93\% \text{ uncertainty reduction about Y by knowing X}$$

$$t(Y;X) = 65\% \text{ uncertainty reduction about X by knowing Y}$$

$$H(X/Y) + H(Y/X) + T(Y;X) = H(X,Y)$$

In this example, 93% of our uncertainty about Y is removed by
knowing X. This is so because knowing the row of an entry
completely determines which column it is in except for one entry. If we are told that $b_1 c_2$ occurred, we are not completely certain about the next event. It could be either a_2 or c_2.
The relationship shown in the last line of the table of infor-

TABLE 9

*Matrix Representation of Behavioral Act Sequence
Involving Three Animals.   See Text for Explanation of Method.*

$$a_1\ b_2\ c_1\ b_2\ b_1\ a_2\ b_1\ c_2\ a_2\ b_1\ c_2\ c_1\ b_2\ \cdots$$

Y

X		a_1	a_2	b_1	b_2	c_1	c_2
$a_1\ a_1$ / a_2 / b_1							
	b_2					1	1
c_1 / c_2 / $a_2\ a_1$ / a_2							
	b_1					2	2
b_2 / c_1 / c_2 / $b_1\ a_1$							
	a_2		1				1
b_1 / b_2 / c_1							
	c_2		1			1	2
$b_2\ a_1$ / a_2							
	b_1		1				1
	b_2						
	c_1				1		1
c_2 / $c_1\ a_1$ / a_2 / b_1							
	b_2			1			1

continued on following page

	a_1	a_2	b_1	b_2	c_1	c_2
c_1						
c_2						
$c_2 a_1$						
a_2			1			1
b_1						
b_2						
c_1				1		1
c_2					1	1
Totals:		2	3	2	2	3

mation values states that "the information about X which remains after Y is known, plus the information about Y which remains after X is known, plus the shared information of both X and Y, is equal to the total information in the two characters". This relationship has been discussed by Estabrook (1967, In Press).

A.   Statistical Confidence Limits and Hypothesis Testing

Until now, a major drawback of information theory has been the absence of traditional confidence limits for the generated values. This problem has been solved independently by Drs. George Losey (personal communication) and Robert Fagen (In Press). In his invaluable report on this subject, Fagen evaluated two methods of estimating variance of information measures and of reducing the bias of such measures. One method involves a "Jacknife technique" and the other is a closed-form approximation. His results indicate that the closed-form procedures are preferable to the jacknife and this discussion will therefore be confined to them.

B.   Bias Correction

It has been frequently pointed out that information measures calculated from samples of behavior are subject to systematic errors (e.g., Altmann, 1965; Chatfield and Lemon, 1970; Hazlett and Estabrook, 1974a,b). To correct for first-order bias the following expressions can be used (Fagen, In Press):

$$H(X) = \hat{H}(X) + \frac{r-1}{(1.3863)N}$$

$$H(Y/X) = \hat{H}(Y/X) + \frac{r(c-1)}{(1.3863)N}$$

$$T(X;Y) = \hat{T}(X;Y) - \frac{(r-1)(c-1)}{(1.3863)N}$$

where r = the number of rows in a contingency table = the number of states in X, c = the number of columns = the number of states in Y, and N = the total of all observations of all the states of X and Y. According to these expressions, H(X) and H(Y/X) tend to be underestimated, while T(X;Y) tends to be overestimated. Returning to the sample data in Table 4 we can easily calculate these bias reduction terms. $\hat{H}(X)$ bias corrected = 1.5219 + 0.0144 = 1.5363 bits; $\hat{H}(Y)$ bias corrected = 1.2955 + 0.0144 = 1.3099 bits; $\hat{H}(Y/X)$ bias corrected = 0.9245 + 0.0432 = 0.9677 bits; and $\hat{T}(X;Y)$ bias corrected = 0.371 - 0.0288 = 0.3429 bits. According to Fagen the application of these correction terms to the measures reduces the bias by one order of magnitude; residual bias is ten times lower than the original raw score.

C. Estimates of Variance

The closed form estimates of variance are as follows (Fagen, In Press):

$$\text{Var } \hat{H}(X) = [\sum_{i=1}^{r} (\hat{P}i(\log_2\hat{P}i)^2) - \hat{H}(X)^2]/N$$

$$\text{Var } \hat{H}(Y/X) = [\sum_{i=1}^{r} \sum_{j=1}^{c} (\hat{P}ij(\log_2\frac{\hat{P}ij}{\hat{P}i})^2) - \hat{H}(Y/X)]/N$$

$$\text{Var } \hat{T}(X;Y) = [\sum_{i=1}^{r} \sum_{j=1}^{c} (\hat{P}ij(\log_2\frac{\hat{P}ij}{\hat{P}i\ \hat{P}j})^2) - \hat{T}(X;Y)]/N$$

Referring again to Table 4 we can calculate the variances.

$$\text{Var } \hat{H}(X) = [(0.4(\log_2 0.4)^2 + 0.2(\log_2 0.2) + 0.4(\log_2 0.4)^2)$$
$$- (1.5363)^2]/100 = [2.4763-(1.5363)^2]/100 = 0.0012$$

$$\text{Var } \hat{H}(Y) = [(0.3(\log_2 0.3)^2 + 0.6(\log_2 0.6)^2 + 0.1(\log_2 0.1)^2)$$
$$- (1.3099)^2]/100 = [2.3345-(1.3099)^2]/100 = 0.0052$$

$$\text{Var } \hat{H}(Y/X) = [(0.1(\log_2\frac{0.1}{20.4})^2 + 0.3(\log_2\frac{0.3}{20.4})^2 + 0.2(\log_2\frac{0.2}{20.4})^2$$

$$+ 0.2(\log_2\frac{0.2}{20.4})^2 + 0.1(\log_2\frac{0.1}{20.4})^2 + 0.1(\log_2\frac{0.1}{20.4})^2]$$

$$- (0.9677)^2]/100 = (1.4516-(0.9677)^2)/100 = 0.0052$$

$$\text{Var } \hat{T}(X;Y) = 0.1(\log_2\frac{0.1}{(0.4)(0.3)})^2 + 0.3(\log_2\frac{0.3}{(0.4)(0.6)})^2$$

$$+ 0.2(\log_2\frac{0.2}{(0.2)(0.6)})^2 + 0.2(\log_2\frac{0.2}{(0.3)(0.4)})^2$$

$$+ 0.1(\log_2\frac{0.1}{(0.6)(0.4)})^2 + 0.1(\log_2\frac{0.1}{(0.4)(0.1)})^2$$

$$= [0.5894 - (0.3429)^2]/100 = 0.0047$$

The need for caution is great since we do not immediately know the underlying distribution of these parameters, although with sufficient sample sizes they may approach a normal distribution.[1] Once these variances have been calculated, we can cautiously use them for hypothesis testing. For example a standard t test applied to test the null hypothesis that H(X) = H(Y) results in a t statistic of 10.96 with 4 degrees of freedom ($p < 0.001$) and the null hypothesis is rejected.

When these expressions were applied to the data of Steinberg and Conant (1974) the variances were all extremely low. For example the variances of H(A) and H(B) in the A-B direction were 0.001 and 0.0003, respectively. Variances of H(B) and H(A) in the B-A direction were 0.0004 and 0.0001 respectively; variance of H(B/A) in the A-B direction was 0.0003 and the variances of T(A;B) in the A-B direction and T(B;A) in the

1. Editor's footnote: The nature of the distribution of information measures depends in a complex way on the conditional probability distributions and the extent of dependency between the distributions (i.e., the apparent transmission rate). Simulations of the distributions of information measures by Dr. William DuMouchel and his students at the University of Michigan (utilizing a slightly different derivation of the variances), indicate that in most cases the significance of the information values can be tested by comparison via a non-central chi-square (Chapter 28 in N.I. Johnson and S.K. Kots (1970) Continuous Univariate Distributions, Vol. 2, Houghton Mifflin Co. Boston.) The expected distributions of such noncentral chi-squares are being generated.

B-A direction were 0.00008 and 0.0002. With such low variances almost any difference between the measures will be a
significant difference. For example a t test of the null
hypothesis that $T(A;B) = T(B;A)$ resulted in a t statistic of
85.02, 22 degress of freedom, $p \ll 0.005$. It follows that the
conclusions of these authors regarding the differences between
responders and initiators of social interactions are statistically valid statements.

Fagen (In Press) warns that these measures tend to become
unreliable when the transmissions are high, and cautions
against their use under such circumstances. As he points out,
however, the transmissions that have been measured in communication studies thus far have been low, and under such circumstances there is no problem with closed form procedures. High
transmissions will be found when the behavior of one animal
very closely constrains that of another animal. In such cases
there can be little doubt that the two are communicating, and
these techniques will be unnecessary. Fagen concludes, "it is
only in the case of low, uncertain H_t (transmission), where
communication is not obvious, that there is any real need for
statistical analysis, and fortunately the procedures analyzed
here exhibit their best performance under precisely these conditions."

D. Sample Size

According to Estabrook (In Press), the minimum sample
size for use in character analysis is equal to 3 (number of
states in X) (number of states in Y). Fagen found difficulties with such a small sample size and recommends 10 (repertoire size)^2, $(10(R)^2)$, for communication studies. This is
the same as 10 (number of states in X) (number of states in Y).
Since Fagen found such a sample size sufficient for dependable
results in his evaluations, it is probably unnecessary to
collect as much data as has been recommended and attempted in
the past (e.g., Steinberg and Conant (1974) used a sample size
slightly larger than $50(R)^2$). Therefore, informational
analysis of dyadic interactions should become a much more useful tool to fit the needs of shorter term experiments. If,
however, one intends to deal with multi-individual interactions or three or more characters at a time, sample size again
becomes a problem. This is because the number of possible
events is at once magnified. Our mythical animal with its 3
act repertoire can be handled quite nicely with $10(R)^2$ observations when we are dealing with two animals interacting.
When three interact we need $10(R)^3$ observations and when there
are four interactors $10(R)^4$. It is easy to see that in dealing with animals with repertoires of 10-15 acts we are going

to need a great deal of data, indeed, to handle more than two animals or variables at one time. Bear in mind, however, that many combinations are impossible even with dyadic interactions (e.g., "touch" cannot follow "retreat"). With more than two variables to consider the number of impossible transitions will be even higher and these should not be counted when determining sample size. Count the cells in the contingency table, subtract those cells for which there can be no entries (impossible transitions) and multiply the result by ten. This will give a reasonable approximation to the total number of entries the table should contain before calculating estimates of information theoretic measures.

CONCLUSIONS

In the foregoing discussion I have tried to summarize and synthesize the various information theoretic formulations which can be applied to ethological data. I have included examples of how these measures are calculated, so that the mathematically unsophisticated reader can learn to use these valuable tools. No model is perfect and this one is not offered as a panacea for all problems in sequential analysis. The problem of non-stationarity of the probabilities used in calculation of these measures is frequently cited as a reason for not using them at all. This problem has several solutions. First of all, when the measures are calculated on such characters as positions in the sequence (i.e., all first acts, second acts, etc.), the probabilities are basically stationary. Furthermore, when in doubt, there are standard tests for stationarity in the statistical literature and the amount that the information theory measures become inflated due to non-stationarity can be subtracted from the estimated values.

With a sufficiently large sample size (but not as vast as formerly advocated), the formulations offer an intuitively satisfying way of investigating constraints between variables. I have found the results both practical and informative, and hope that others will find them equally so.

ACKNOWLEDGEMENTS

I am, as always, extremely grateful to Roger Conant for his critique of this paper and for all of his ideas about the ways in which information theory can be applied to behavioral problems. It was he who first suggested many of the techniques described herein and I hope I have done them justice. This manuscript was much improved by the suggestions of Robert B. Willey, whose cousel is ever invaluable, and whose virtues as a teacher and honorable human being will always be an inspiration to me. I also wish to thank Gary Hyatt for his helpful comments at all stages of this undertaking.

REFERENCES

Altmann, S.A. 1965. Sociobiology of rhesus monkeys, II. Stochastics of social communication. J. Theor. Biol. 8: 490-552.
_____. 1967. The structure of primate social communication. In Social Communication among Primates, S.A. Altmann (ed.), Univ. Chicago Press, Chicago.
Baylis, J. 1975. A quantitative, comparative study of courtship in two sympatric species of the genus Cichlasoma (Teleostei, Cichlidae). Ph.D. Dissert., Univ. of California at Berkeley.
Chatfield, C. and R.E. Lemon. 1970. Analysing sequences of behavioural events. J. Theor. Biol. 29: 427-455.
Conant, R. and J.B. Steinberg. 1973. Information exchanged in grasshopper interactions. Info. & Control. 23: 221-233.
Dingle, H. 1969. A statistical and information analysis of aggressive communication in the Mantis Shrimp Gonodactylus bredini Manning. Anim. Behav. 17: 561-575.
_____. 1972. Aggressive behavior in stomatopods and the use of information theory in the analysis of animal communication. p. 126-156 in H.E. Winn and B.L. Olla (eds.) Behavior of Marine Animals, Vol. 1 Invertebrates, Plenum Press, New York.
Estabrook, G.W. 1967. An information theory model for character analysis. Taxon. 16: 86-97.
_____. In Press. An Introduction to Environmental Measurement and its Descriptive Analysis by Computer. David McKay, New York.
Fagen, R.M. In Press. Information measures: Statistical confidence limits and inference. J. Theor. Biol.
Hazlett, B.A. and W.H. Bossert. 1965. A statistical analysis of the aggressive communications systems of some hermit crabs. Anim. Behav. 13: 357-373.
_____ and G.F. Estabrook. 1974a. Examination of agonistic behavior by character analysis. I. The spider crab Microphrys bicornutus. Behaviour 48: 131-144.
_____. 1974b. Examination of agonistic behavior by character analysis II. Hermit crabs. Behaviour 49: 88-110.
Hyatt, G.W. In Preparation. A comparative information theory analysis of aggression in two species of fiddler crabs (Uca pugilator and U. pugnax).
Manning, A. 1976. The place of genetics in the study of

behaviour. p. 327-343 in Growing Points in Ethology.
P.P.G. Bateson and R.A. Hinde, (eds.) Cambridge Univ.
Press, Cambridge.

Rubenstein, D.I. and B.A. Hazlett. 1974. Examination of the
agonistic behaviour of the crayfish Orconectes virilis
by character analysis. Behaviour 50: 193-216.

Shannon, C.E. and W. Weaver. 1949. The Mathematical Theory
of Communication. Univ. of Illinois Press, Urbana.

Steinberg, J.B. and R.C. Conant. 1974. An informational
analysis of the inter-male behaviour of the grasshopper
Chortophaga viridifasciata. Anim. Behav. 22: 617-627.

SPIDERS & SNAILS & STATISTICAL TALES:
APPLICATION OF MULTIVARIATE
ANALYSES TO DIVERSE ETHOLOGICAL DATA

Wayne P. Aspey & James E. Blankenship

The University of Texas

Abstract: Multivariate analyses organize and reduce data composed of numerous variables into fewer biologically inter-pretable dimensions. This paper describes the usefulness of R- and Q-type orthogonal factor analysis, linear typal analysis, and other classification typologies (e.g., multidimensional scaling, principal-components analysis, cluster analysis) in uncovering homogeneous subgroups from naturally-selected, heterogeneous samples of animals or behaviors. Once animal groups are identified according to some criteria, multiple step-wise discriminant analysis can be used to determine which behavioral variables best differentiate between all group com-binations. R-type factor analysis applied to 20 agonistic behaviors exhibited during adult male-male agonistic interac-tions in the wolf spider Schizocosa crassipes yielded four behavior-related factors: (I) Approach/Signal, (II) Vigorous Pursuit, (III) Run/Retreat, and (IV) Non-Linking Behaviors. If the same matrix is rotated and a Q-type factor analysis is applied (this time to the 40 subjects as variables), two subject-related factors were extracted and interpreted as (I) Dominance and (II) Subordinance. Discriminant analysis of spider groups pre-identified on the basis of density para-meters of dominance rankings characterized the groups in terms of the original 20 behaviors and indicated which behaviors most optimally discriminate between all pairwise group combin-ations. Q-type orthogonal powered-vector factor analysis and linear typal analysis of 10 burrowing parameters from the marine gastropod mollusc Aplysia brasiliana yielded three fac-tors interpreted as Efficient, Inefficient, and Intermediate burrowers. Used as a diagnostic tool in proper perspective, multivariate analyses structure complex data, provide insight into underlying dimensions and sources of individual varia-tion, and facilitate the formulation of testable hypotheses for studying mechanisms controlling behavior.

INTRODUCTION

The animal behaviorist's task in analytically descriptive ethology is to describe meticulously a species' behavioral repertoire in operational terms, and to determine the inherent structure of the data in terms of underlying relationships between/among the variables. Once this underlying organizational structure is uncovered, a variety of experimental advantages can result: (1) additional variables will suggest themselves as important in influencing behavior; (2) insight will be gained into the behavior's functional significance; and (3) meaningful and testable hypotheses can be formulated to guide future research.

A central problem in ethology focuses on the selection, application, and interpretation of appropriate analytical techniques for dealing with life's variability (including an animal's behavior as well as the investigator's approach). As Marc Bekoff aptly stated in another of this volume's chapters: "... fit the method of analysis to the animal, not the animal to the method..."; permit me to extend this suggestion to include "... and the problem under investigation." Just as every animal exhibits variability in terms of unique morphological and behavioral characteristics and limitations, each investigator selects problems, designs and executes experiments, and measures behavior with as much diversity.

Therefore, in this chapter I hope to illustrate the usefulness of multivariate statistical techniques for: (1) uncovering structure (i.e., decreasing variability) within a heterogeneous sample without reference to a pre-existing classification scheme; and (2) for hypothesis-seeking once some systematic order becomes apparent. To this extent, I shall survey examples from my own research on agonistic behavior in adult male wolf spiders (Aspey, In Press[a,b]) and burrowing in marine snails (Aspey and Blankenship, 1976a,b,c, In Review) in an effort to demonstrate how results obtained from multivariate statistics are useful in achieving the three experimental advantages mentioned in the first paragraph.

Multivariate analyses are useful for the organization, analysis, and interpretation of data involving numerous subjects with measurements on multiple variables. In contrast to experimentally-oriented behavioral research which focuses on a single dependent variable while holding extrinsic sources of variation constant, ethologically-oriented research is aimed more toward observing freely-behaving animals under "naturalistic" circumstances. Consequently, ethological data tend to be multidimensional, widely variable, scaled in arbitrary units, and frequently include interacting variables. Thus, control over important sources of variation is often impossible, especially when: (1) animals selected for observation are sampled from a population composed of several distinct subpopulations (e.g., dominant and subordinate animals); and (2) numerous behavioral measurements are recorded on each subject (e.g., frequency counts of behaviors exhibited during agonistic interactions).

In essence, multivariate analyses are potentially powerful diagnostic tools: (1) for uncovering homogeneous subgroups from naturally-selected heterogeneous samples; and (2) for identifying relationships among multiple variables when the underlying source, or biological basis, of individual variation is unknown. When observations are made on multiple variables, certain modal patterns recur frequently, and it is inferred that these patterns represent homogeneous subtypes. Conceptually, this idea is familiar in ethology, given the historically extensive literature on "Fixed Action Patterns" and Barlow's (1968) contemporary proposal of "Modal Action Patterns" (also Bekoff, this volume). Since understanding relationships among numerous variables requires treatment more comprehensive than afforded by traditional statistics, multivariate statistics provide useful alternatives (Morrison, 1967; Hope, 1968; Cooley and Lohnes, 1971; Overall and Klett, 1972; Atchley and Bryant, 1975a,b; Bishop, Fienberg, and Holland, 1975). Furthermore, multivariate analyses allow the visualization of numerous variables simultaneously through numeric, geometric, and graphic expression (Rohlf, 1968).

Multivariate techniques minimize the number of variables by classifying individuals and/or examining similarities and differences among them in terms of multiple measurements. Such empirical classification typologies (e.g., factor analysis, linear typal analysis, principal-components analysis, multidimensional scaling, cluster analysis) are derived from internal relationships among the variables without reference to any pre-existing classification system. Therefore, measurement variables used for defining similarities and differences among individuals should be relevant. However, this is not always possible at the onset of ethological data collection, and multivariate techniques assist in providing insight into the relevance of certain behavioral variables that may not have been obvious initially.

Multivariate analyses have been widely used in both plant and animal ecology (Williams and Lambert, 1959; Cassie and Michael, 1968; Cassie, 1969; James, 1971), systematic zoology (Rohlf and Sokal, 1958; Rohlf, 1968, 1971; Rohlf and Kispaugh, 1972; Leamy, 1975), and psychology (Ruskin and Corman, 1971; Royce, Poley, and Yeudall, 1973; Ekehammar, Schalling, and Magnusson, 1975), but their application in animal behavior is relatively novel (Dudziński and Norris, 1970; Morgan, Simpson, Hanby, and Hall-Craggs, 1975; Huntingford, 1976b). Since Wiepkema (1961) introduced factor analysis to ethology, animal behaviorists have only recently begun to use multivariate techniques as an aid in the experimental analysis of behavior. To illustrate, behavioral studies demonstrating the power of multivariate analyses for (1) establishing homogeneous subgroups from naturally-sampled populations, and (2) identifying relationships among the extracted subgroups include: behavioral genetics of human twins (Eaves, 1972), human privacy (Marshall, 1972), play in children (Smith and Connolly, 1972), signals in squirrel monkeys (Marcus and Pruscha, 1973), marmot social organization (Svendsen and Armitage, 1973), bee colony interactions (Brothers and Michener, 1974), canid behavioral taxonomy (Bekoff, Hill, and Mitton, 1975, and references in Bekoff, this volume), grooming in flies (Dawkins and Dawkins, 1976), newt sexual behavior (Halliday, 1976), territorial aggression in sticklebacks (Huntingford, 1976a,b), cat neuroethology (Schwartz, Ramon, and John, 1976), burrowing in marine snails (Aspey and Blankenship, 1976a,b,c), wolf spider sociobiology (Aspey, In Press$_{a,b}$), and fiddler crab behavioral ecology (Aspey, In Press$_c$).

Although I advocate quantitative approaches to ethology, allow me to caution that multivariate analyses are simply one research tool available to the animal behaviorist for determining inherent data structure. Just as the "Skinner Box" revolutionized the experimental analysis of learning, the judicious use of multivariate statistical techniques can also revolutionize the analytically descriptive aspects of ethology. However, misusing multivariate analyses in ethology could most certainly lead to a bewildering array of extraordinarily sterile papers. Instead, used in proper perspective as a diagnostic tool for gaining insight into underlying dimensions of individual variation, multivariate analyses can help generate meaningful and testable hypotheses for guiding future research aimed at uncovering mechanisms controlling behavior.

I. SPIDERS...

In spite of the extensive literature on agonistic behavior in arthropods (Dingle and Caldwell, 1969; Hazlett, 1968, 1972, 1974; Ewing, 1972; Hazlett and Estabrook, 1974a,b), the spiders remain neglected. Of the existing spider behavior studies, the majority deal with aspects of web-building (Witt, Reed, and Peakall, 1968; Eberhard, 1969) or reproductive behavior (Crane, 1949; Rovner, 1966, 1967, 1968, 1971, 1972, 1974). This void in the literature may be due to: (1) fear

TABLE 1

*Behavioral Repertoire of Adult Male Schizocosa crassipes
During Agonistic Interactions*

I. Locomotory Behaviors

  A. Approach Behaviors
1. Mutual Approach
2. Front Approach
3. Lateral Approach
4. Posterior Approach
5. Diagonal Approach
6. Following Walk
7. Chase

  B. Avoidance Behaviors
1. Mutual Avoid
2. Retreat
3. Run

II. Contact

III. Foreleg Movements and Postures

  A. Ipsilateral Foreleg Movements and Postures
1. Wave and Arch
2. Prolonged Wave

  B. Synchronous Foreleg Movements and Postures
1. Tapping
2. Jerky Tapping
3. Horizontal Extend
4. Oblique Extend
5. Vertical Extend
6. Up-Down
7. Vibrate-Thrust

of and/or the "maligned reputation" of spiders; (2) the idea that most spiders are "solitary" and/or asocial; or (3) the comments of early workers that spiders simply cannabalize or ignore one another and do not fight (Montgomery, 1910; Bristowe, 1929). Field and laboratory observations of the brush-legged wolf spider Schizocosa crassipes Walckenaer quickly dispel these unfounded misconceptions (see also Buskirk, 1975; Burgess, 1976).

These small spiders (leg span approximately 2 cm) are harmless and show strikingly beautiful sexual dimorphism, the males developing conspicuous tufts of black hairs on the tibiae and patellae of the forelegs at the final molt. This species is found in dense abundance in circumscribed areas at the forest-meadow interface of deciduous forests (Aspey, 1976a), and field and laboratory observations revealed an elaborate behavioral repertoire that occurred exclusively during adult male-male agonistic interactions. Twenty behaviors were reliably identified, operationally defined, analyzed by cinematographic techniques, and descriptively categorized as: (1) locomotion: approach or avoidance; (2) contact; and (3) movements and/or postures of one or both forelegs (Table 1). Figure 1 illustrates several of the foreleg movements and/or postures.

Fig. 1. Foreleg movements and/or postures observed during agonistic interactions between adult male Schizocosa crassipes.

(A) <u>Resting Stance</u>. Measurements of foreleg movements and/or postures were determined relative to the Resting Stance, in which forelegs are extended forward with the patellae flexed downward and the tarsi contacting the substratum.

(B) <u>Wave and Arch</u>. Wave and Arch occurred when a foreleg was lifted rapidly once or twice in succession through a 50-60° arc. Each wave was accompanied by an arch formed by bending the tibio-metatarsal joint 65-80°. Wave and Arch appeared to be a low intensity threat display from which conspecifics retreated.

(C) <u>Prolonged Wave</u>. Prolonged Wave was a series of slow single foreleg waves each lasting 3 to 4 sec and repeated at least three consecutive times. The configuration of the wave was the same as in Wave and Arch, however, Arch accompanied Prolonged Wave less than one-third of the time. Prolonged Wave appeared to be an appeasement display.

(D) <u>Horizontal Extend</u>. During Horizontal Extend (spider on the right), all foreleg segments were straightened and held horizontally and parallel to one another and the substratum. Conspecifics responsed to Horizontal Extend by Retreat (65% of the time) or Run (7.4% of the time).

(E) <u>Oblique Extend</u>. During Oblique Extend both forelegs are extended 45° relative to the substratum with all leg segments fully straightened. Oblique Extend also appeared to function as a low intensity threat display.

(F) <u>Vertical Extend</u>. In Vertical Extend the femora of both forelegs were held 70 to 100° relative to the substratum with the abdomen lowered. Vertical Extend, Run, and Retreat were characteristic of Subordinate spiders, and these behaviors were exhibited in response to actively approaching conspecifics.

(G) Up-Down. Up-Down was performed by alternately elevating and lowering the forelegs three or four times so that one leg was extended "up" (similar to Oblique Extend) while the other was extended "down" (similar to the Resting Stance). Up-Down signalled subsequent mutual avoidance by conspecifics.
(H) Vibrate-Thrust. Vibrate-Thrust (spider on the left) was the most vigorous agonistic behavior displayed in which the forelegs were rapidly moved back and forth several times before the spider lunged upward and forward toward its opponent. Spiders run away in response to Vibrate-Thrust. [Figure reprinted from Aspey (In Press$_a$) with permission from Behaviour].

A. Determining Structure of the Behaviors

To determine whether the 20 behaviors observed during 3,525 agonistic interactions (14,109 behavioral acts) by 40 adult male S. crassipes could be associated together in some meaningful pattern, an R-type orthogonal principal-axis factor analysis with Kaiser's Varimax rotation was applied to the data.

Factor Analysis as a General Method

Factor Analysis (Rummel, 1970; Comrey, 1973) is one strategy for reducing large correlated categorical data into a smaller number of uncorrelated factors and it aims to determine the structure among multiple variables by showing that a relatively few factors account for a large proportion of the variance among the original variables. These factors, then, suggest some hypothesis concerning the underlying relationships among the multiple variables. Factors are considered primary dimensions of individual difference, each representing a distinctly different source of variation. Desirable properties for a factor solution include: (1) parsimony in the number of factors explaining most of the variance among multiple variables; (2) orthogonality, or at least relative statistical independence, of the factors; and (3) conceptual meaningfulness of the factors as they relate to the variables (Comrey, 1973).

R-Factor Analysis

In this application with spiders, R-factor analysis (Rummell, 1970) treats the 20 behaviors, with corresponding scores from 40 subjects, as the variables. Behavior-related factors are extracted from a data matrix of 20 behaviors (columns) x 40 spiders (rows) to determine whether the 20 behav-

iors can be organized into a fewer number of underlying factors. In contrast, Q-factor analysis treats the subjects,
with their corresponding scores on the 20 behaviors, as the
variables and extracts subject-related factors from a matrix
of 40 spiders (columns) x 20 behaviors (rows). Q-factor
analysis will be discussed later.

To compute an R-factor analysis, data are cast into a
"behaviors x subjects" matrix and a product-moment correlation
coefficient matrix is computed. Eigenvalues are extracted
from the correlation matrix, but only those eigenvalues $\geq$ 1.0
are retained for determining the number of factors. Eigenvalues (or latent roots) are proportional measures of the total variance accounted for by successively extracted factors
(i.e., Factor 1 accounts for the largest proportion of
variance, etc). Orthogonal rotation using Kaiser's Normal
Varimax method (Kaiser, 1958, 1959) was then performed on the
factor matrix. Kaiser's Varimax rotation is a procedure almost always required to establish meaningful relations between factors and the original variables following principal-
axis factor analysis (Overall and Klett, 1972). Simply, factors extracted using the principal-axis method are statistically orthogonal and account for maximum possible variance
among multiple variables. Comrey (1973) provides an under-
standable textbook account of factor analysis for the novice,
while Rummel (1970) and Overall and Klett (1972) provide more
advanced treatments.

R-factor analysis extracted four behavior-related factors
which accounted for 74.3% of the total variance (Table 2).
Numbers under the four "Extracted Factors" columns are "factor
loadings" which represent the extent each behavior correlates
with the extracted factors. Although factor loadings as low
as 0.30 are commonly reported, such values are too low for
reliable interpretation (Comrey, 1973). As a guide, Comrey
(1973) suggests and rates the following "zero factor loadings"
(i.e., lower cut off values for indicating factor loadings)
for orthogonal factor interpretation: 0.71 -- excellent;
0.63 -- very good; 0.55 -- good; 0.45 -- fair; and 0.32 --
poor. To determine the percent of variance in common between
a given variable and the factor, square the factor loading.
In Table 2 , for example, Up-Down has a factor loading of 0.89
on Factor 1 and shares 79% of the variance in common with that
factor $[(0.89)^2 = 0.79 = 79\%]$. On this basis, 0.45 was considered the zero factor loading for this study (i.e., factor
loadings less than 0.45 are not listed in Table 2) since such
variables would share less than 20% of the variance in common
with the factor. Furthermore, only factor loadings $\geq$ 0.55 are
considered in factor interpretation (variable and factor
sharing 30% of the variance in common).

TABLE 2

Four behavior-related factors and accompanying factor loadings extracted from a Varimax rotated R-type orthogonal principal-axis factor analysis employing 20 behaviors as variables. The behaviors were exhibited by 40 adult male Schizocosa crassipes during agonistic interactions. Zero factor loading was 0.45 throughout. [Data modified from Aspey (In Press$_a$) with permission from Behaviour].

Behavior Number	Behavior Description	Extracted Factors			
		I	II	III	IV
19	Up-Down	*0.89*			
16	Horizontal Extend	*0.84*			
17	Oblique Extend	*0.79*			
3	Lateral Approach	*0.76*	0.46		
12	Wave and Arch	*0.74*			
2	Front Approach	*0.66*	0.49		
5	Posterior Approach	*0.65*	0.52		
8	Mutual Avoid	*0.63*			
11	Contact	*0.63*			0.51
20	Vibrate-Thrust		*0.89*		
7	Chase		*0.88*		
15	Jerky Tapping	0.50	*0.75*		
6	Following Walk	0.51	*0.74*		
1	Mutual Approach		*0.73*		
10	Run			*0.82*	
18	Vertical Extend			*0.74*	
9	Retreat	0.51		*0.57*	
14	Tapping				*0.81*
4	Diagonal Approach				*0.64*
13	Prolonged Wave				*0.61*
Eigenvalues..................		9.5	2.6	1.7	1.2
Variance explained...........		47.4%	12.8%	8.3%	5.8%
Cumulative variance..........		47.4%	60.2%	68.5%	74.3%

Factor Interpretations

To interpret the structure among the multiple variables as determined by R-factor analysis, the four extracted factors were descriptively named and assessed according to the behaviors comprising them. For example, Factor I accounted for 47.4% of the variance, consisted of behaviors involving approaches toward conspecifics, contact, and various foreleg movements and/or postures, and was labelled "Approach/Signal". Factor II accounted for 12.8% of the variance and was labelled "Vigorous Pursuit" since the five behaviors comprising this factor involved a vigorous approach and pursuit of conspecifics. Factor III accounted for 8.3% of the variance and was termed "Run/Retreat" since Run, Retreat, and Vertical Extend comprised this factor. The groupings of Vertical Extend with Run and Retreat provided insight into the significance of Vertical Extend since this posture not only occurred in response to one spider's being approached or pursued by a conspecific, but also was strongly associated with retreating animals when intra-individual behavior sequences were analyzed (Aspey, 1976b). Finally, Factor IV accounted for 5.8% of the variance and at first seemed biologically uninterpretable. However, the common element among these behavior was that none were significantly linked beyond chance expectation with any other behavior during inter-individual interactions (Aspey, In Press$_a$). Since all other behaviors were significantly linked to one or more behaviors during male-male agonistic encounters, Factor IV was labelled "Non-Linking" behaviors. The kinematic diagram represented by Figure 2 graphically structures the behaviors as the R-factor analysis organized them.

Thus, R-factor analysis reduced the 20 original behaviors to four factors, each representing a different source of variation, and this organization suggested some form of communication system among the spiders with the 20 behaviors serving as the animal's "vocabulary" during agonistic encounters. Since each factor represents a different and independent source of variation, a logical question to consider next would be: "Do specific spiders consistently exhibit 'Run/Retreat' behaviors, while others consistently exhibit 'Approach/Signal' or 'Vigorous Pursuit' behaviors?" This consideration leads to one of the strongest advantages of multivariate analyses -- the generation of meaningful and testable hypotheses for guiding further research. Since there appears to be some structure within the data in terms of independent agonistic behavior categories, perhaps a similar ordering among the subjects exists such that the question posed above can be rephrased: "Does a dominance hierarchy exist among spiders?".

B. Determining Structure Among the Spiders

Two possible approaches to answer this question using the same raw data as for the R-factor analysis of the behaviors include: (1) develop a dominance index for each spider and determine if the spiders can be ranked in a dominance hierar-

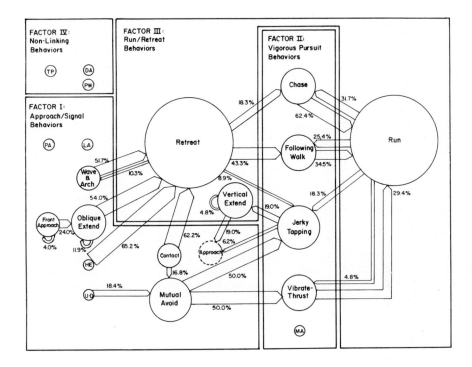

Fig. 2. Twenty behaviors exhibited during 3,525 agonistic interactions by 40 adult male Schizocosa crassipes *are shown organized according to a Varimax rotated R-type orthogonal principal-axis factor analysis. Arrows connecting two behaviors represent significant linkages; relative thickness of the arrows indicate the percent of time one behavior followed another. The relative size of each circle represents the total number of behaviors significantly linked with that behavior. Approach, represented by a dotted outline, was a newly created category derived from combining Mutual Approach, Lateral Approach, Posterior Approach, and Diagonal Approach. Abbreviations: DA = Diagonal Approach, MA = Mutual Approach, PA = Posterior Approach, PW = Prolonged Wave, TP = Tapping, UD = Up-Down. [Figure reprinted from Aspey (In Press$_a$) with permission from* Behaviour].

chy as indicated by Landau's (1951a,b, 1953) index of linearity "h" values (Chase, 1974 and Bekoff, this volume); and (2) determine whether groups corresponding to dimensions of "Dominance" and "Subordinance" would result from applying Q-factor analysis to the data.

The "Dominance Index"

A "Dominance Index" (DI) was developed for each spider on the basis of the kinds, intensity, and frequency of agonistic behaviors exhibited during interactions with all possible partners. Furthermore, given the ethologically-oriented approach of analyzing freely-behaving, spontaneously interacting spiders without using the Round-Robin pairings, the total number of interactions also entered into the DI formula to make DI's comparable from spider to spider. Details regarding the mechanics and other aspects of the DI are provided in Aspey, In Press[a]. Landau's index indicated a perfect linear hierarchy (h=1) that remained stable over 10 days of testing within each test group. For example, one spider (Alpha) in each group had higher DI's over all other spiders in the group; the next spider (Beta) had higher DI's over all other animals in the group except the Alpha spider, and so on. For convenience, those one or two spiders (depending on the number in the group) having the highest DI's in each group were designated "Dominant"; those one or two spiders with the lowest DI's were designated "Subordinate"; and those spiders having as many animals dominant over them as subordinate to them were designated "Intermediate".

Q-Factor Analysis

A second strategy for investigating whether a dominance structure existed among the spiders would be to apply Q-factor analysis to the subjects and determine if the resulting subject-related groups corresponded to the spiders' DI designations. Since animal behaviorists are frequently interested in studying individual differences, Q-factor analysis is a preferred approach for studying natural groupings among subjects (Rummel, 1970; Overall and Klett, 1972; Aspey and Blankenship, 1976a; Aspey, In Press[a]). Therefore, Q-type orthogonal principal-axis factor analysis was applied to the original raw data. In this case, the subjects, with their corresponding scores on the 20 behaviors, were treated as the variables and subject-related groups were extracted. As with R-factor analysis, only factors with eigenvalues ≥ 1.0 were used to determine the number of factors, Kaisers Varimax rotation was used, 0.45 was considered the zero factor loading, and factor interpretation was based on factor loadings ≥ 0.53.

Factor Interpretation

Four subject-related factors accounting for 90% of the variance were extracted, and interpreted by comparing the

TABLE 3

*Four subject-related factors and accompanying factor loadings
extracted from a varimax rotated Q-type orthogonal principal-
axis factor analysis employing 40 adult male Schizocosa
crassipes as variables with scores on 20 behaviors exhibited
during agonistic interactions.   Zero factor loading was 0.45
throughout.   D = Dominant; I = Intermediate; S = Subordinate.
[Data modified from Aspey (In Press$_a$) with permission from
Behaviour].*

Spider Number	Rank by Dominance Index	Extracted Factors			
		I	II	III	IV
3	S	0.93			
9	S	0.87			
35	S	0.87			
23	S	0.85			
30	S	0.80			
19	S	0.79			
39	S	0.77			
13	S	0.77			
8	S	0.74			
11	S	0.71			
18	I	0.70			
38	S	0.68		0.45	
10	D	0.67	0.56		
15	I	0.66			-0.45
17	S	0.66	0.58		
21	S	0.66			-0.52
20	D	0.66		0.55	
25	I	0.63		0.53	
34	I	0.62	0.47		-0.47
26	I	0.59		0.56	
2	S	0.57	0.54	0.47	
28	S	0.63		0.66	
12	D			0.85	
27	D	0.47	0.48	0.64	
32	S			0.54	
36	D		0.98		
1	D	0.48	0.80		
29	D		0.79		
14	D		0.74		0.48
37	D	0.52	0.72		
7	D		0.71	0.65	
4	I	0.51	0.69		

continued on following page

Number	Rank	I	II	III	IV
40	I	0.51	*0.66*	0.47	
16	D		*0.65*		
6	I	0.63	*0.65*		
5	D	0.46	*0.60*		-0.51
22	D		*0.59*		
31	D	0.51	*0.53*		
24	D		0.48		*-0.64*
33	D	0.51			*-0.62*
Eigenvalues.................		30.3	3.0	1.5	1.2
Variance explained..........		75.8%	7.5%	3.7%	3.0%
Cumulative variance.........		75.8%	83.3%	87.0%	90.0%

spiders associated within each factor with the spider's DI classification (Table 3). Factor I accounted for 75.8% of the variance, and those 10 (48%) subjects with the highest factor loadings ($\geq$ 0.71, Comrey's rating of "excellent") were ranked as Subordinate by the DI. The remaining 11 (52%) spiders with factor loadings 0.57 to 0.70 were predominantly ranked Subordinate (N=4) and Intermediate (N=5). On this basis, Factor I was designated "Subordinance". Factor II accounted for 7.5% of the variance, and of the 13 spiders comprising this factor, 10 (77%) were ranked as Dominant by the DI while the remaining three were ranked Intermediate. Since six of these Dominant spiders had factor loadings $\geq$ 0.71, this factor was designated "Dominance". Factors III and IV accounted for 3.7% and 3.0% of the variance, respectively, and were composed of only six subjects (#28,12,27,32,24,33). Furthermore, since two of these spiders (#28,32) had factor loadings nearly as high on Factor I, and since no common underlying characteristics were apparent among the six spiders, these factors were considered biologically uninterpretable.

Thus, Factor I appeared to represent a continuum of Subordinate and Intermediate spiders with Subordinate spiders having higher factor loadings and Intermediate spiders lower factor loadings on this factor. Similarly, Factor II appeared to represent a continuum of Dominant and Intermediate spiders with Dominant spiders having the higher factor loadings on this factor. Figure 3 illustrates the visual and biologically interpretable representation of this dominance continuum by plotting the factor loadings of the 34 spiders from Factors I and II against one another.

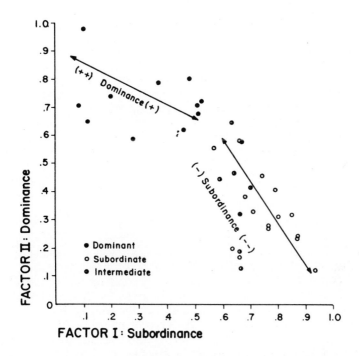

*Fig. 3. Thirty-four adult male <u>Schizocosa</u> <u>crassipes</u>
plotted in the space of orthogonal Factors I and II from a
Varimax rotated Q-type principal-axis factor analysis. Of 21
spiders comprising Factor I, the 10 that received the highest
factor loadings (> 0.71) were independently ranked as Subordinate by the Dominance Index (DI). Of 13 spiders comprising
Factor II, the 10 that were independently ranked as Dominant
by the DI also had the highest factor loadings on this factor.
Spiders were considered Intermediate in rank if there were as
many spiders dominant over them as subordinate to them.*

C. Discriminating Between and Among Groups

Although Q-factor analysis extracted natural groupings of
spiders that corresponded with the DI classification, factor
analysis failed to systematically separate Intermediate from
Dominant and Subordinate spiders. To resolve this apparent
inconsistency, multiple stepwise discriminant analysis was
used (see also Bekoff's discussion of linear function discriminant analysis in this volume). Whereas factor analysis and
other multivariate statistical techniques (discussed in the
STATISTICAL TALES section of this paper) are classification
typologies for identifying homogeneous subgroups from a naturally-selected heterogeneous sample, multiple stepwise discrim-

inant analysis (1) determines relationships among several
identified groups, (2) assesses the discriminability of these
identified groups in terms of measurements on multiple variables, and (3) classifies individual subjects among the several groups. In essence, once different groups are pre-
identified according to some rationale (e.g., by factor
analysis, other classification procedures, or experimental
design contingencies), multiple stepwise discriminant analysis
can verify their distinctness, as well as determine which of
the multiple measurements best discriminates between any two
groups combinations.

Multiple Stepwise Discriminant Analysis

Multiple stepwise discriminant analysis (hereafter referred to as "discriminant analysis") finds a subset of variables (i.e., behaviors) that maximizes group differences
among subjects by reducing multiple measurements (i.e., frequency counts on the 20 behaviors) to one or more combinations
having maximum potential for distinguishing among the different groups (Overall and Klett, 1972; Jennrich, 1977). The
first discriminant function (first canonical variate) is that
particular composite variable on which the sum of squared
differences among group means is maximally great relative to
the within-group variance for the same weighted composite
(Overall and Klett, 1972). In the analysis used here, a
sequence of discriminant equations was computed in a stepwise
manner so that one variable was added to the equation at each
step, and a one-way analysis of variance F-statistic was then
used to determine which variable should join the function
next. The variable added is the one making the greatest reduction in the error sum of squares. In the case of multiple
groups, one weighted combination of scores may distinguish between certain groups but not between others. When this
occurs, additional composites are required to distinguish between groups not separated by the first discriminant function.
The second discriminant function, then, is that weighted composite, which of all possible weighted composites uncorrelated
with the first, provides for maximum average separation among
the groups relative to the within-group variability.
  Spiders designated as Dominant (N=16), Intermediate
(N=8), or Subordinate (N=16) by the DI comprised the identified
groups, and the same raw data (frequency counts of the original
20 behaviors) were utilized as in the previous factor analyses.
The mean scores on two or three discriminant functions can be
taken as coordinate values to locate the groups in a geometric
space of minimum dimensionality. For example, in Figure 4 the
first discriminant function separated the Dominant and Subordinate spiders along the abscissa (accounting for 63.9% of the

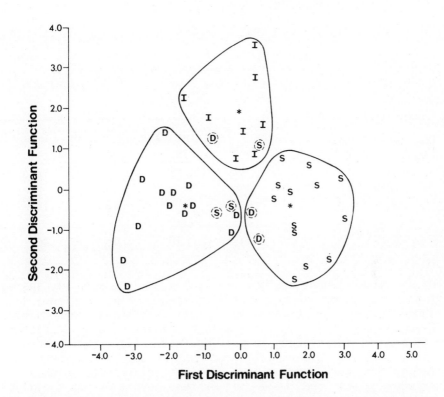

Fig. 4. The locations of 16 Dominant (D), eight Inter-mediate (I), and 16 Subordinate (S) adult male Schizocosa crassipes plotted in a geometric space of minimum dimension-ality by multiple stepwise discriminant analysis on the basis of the frequency of 20 behaviors observed during agonistic en-counters. The first discriminant function (abscissa) is plotted against the second discriminant function (ordinate) and * denotes group means. The spiders were initially grouped as Dominant, Intermediate, or Subordinate by the Dominance In-dex (DI). Dotted circles represent spiders "misclassed" by the DI relative to the discriminant analysis. [Reprinted from Aspey (In Press_a) with permission of Behaviour].

total dispersion), but did not discriminate the Intermediate spiders from either group. Consequently, a second discrimin-ant function was required, which differentiated the Inter-mediate from the Dominant and Subordinate spiders along the ordinate (accounting for 36.1% of the total dispersion).

Classification by Discriminant Analysis

In comparing each spider's dominance classification by

TABLE 4

A comparison of the classification of 40 adult male Schizocosa crassipes into Dominant (DOM), Intermediate (INT), and Subordinate (SUB) groups by the Dominance Index and multiple stepwise discriminant analysis on the basis of 20 behaviors exhibited during agonistic interactions.

Dominance Index	Classification by Discriminant Analysis			
Classification	DOM	INT	SUB	Total
DOM	13	1	2	16
INT	0	8	0	8
SUB	2	1	13	16
Total	15	10	15	40

DI and discriminant analysis, only six spiders were misclassed (Table 4; spiders represented by dotted circles in Figure 4). These spiders corresponded to the six "unclassed" spiders by Q-factor analysis (i.e., those four spiders comprising Factor III and the two spiders comprising Factor IV). Furthermore, all misclassed spiders came from test groups either where only two spiders were present or where five spiders were highly crowded. Thus, under the most extreme conditions of social or spatial density used, the DI was not the best predictor of dominance-subordinance relations. Further influences of varying these density parameters on agonistic behavior in $\underline{S}$. crassipes are presented in Aspey (In Press$_b$).

Figure 4 and Table 4 indicate good within-group clustering of the subjects for each dominance classification, as well as clear discrimination among the groups based on the frequencies of the 20 behaviors. One advantage discriminant analysis has over factor analysis is in the identification of those variables (behaviors) which not only characterize each dominance group, but also differentiate between any two groups. For example, Tables 5, 6, and 7 list those behaviors which optimally differentiate Dominant from Subordinate, Dominant from Intermediate, and Subordinate from Intermediate spiders, respectively. Dominant spiders were characterized by "Approach/Signal" and "Vigorous Pursuit" behaviors, exhibiting a wider behavioral repertoire than Intermediate or Subordinate spiders. Subordinate spiders were characterized by "Run/ Retreat" behaviors, and discriminant analysis indicated that these behaviors, along with Vertical Extend and Prolonged Wave, occurred significantly more in Subordinate than Dominant spiders. Intermediate spiders were also characterized by

TABLE 5

A comparison of the discriminating ability of various behaviors observed during agonistic interactions between Dominant (DOM) and Subordinate (SUB) adult male Schizocosa crassipes. Behaviors are listed in the order in which their frequencies optimally differentiated between spider groups. [Data modified from Aspey (In Press$_a$) with permission from Behaviour].

Behavior	(df)	F-value and Probability Level	
		DOM>SUB	SUB>DOM
Jerky Tapping	(1/37)	7.14**	
Retreat	(2/36)		8.48***
Front Approach	(3/35)	9.08***	
Vertical Extend	(4/34)		7.19***
Oblique Extend	(5/33)	5.65***	
Mutual Approach	(6/32)	4.57**	
Prolonged Wave	(7/31)		4.35**
Contact	(8/30)	4.42**	
Tapping	(9/29)	4.26**	
Diagonal Approach	(10/28)	3.89**	
Wave and Arch	(11/27)	3.42**	
Run	(12/26)		3.27**
Posterior Approach	(13/25)	2.95**	
Mutual Avoid	(14/24)	2.71*	
Vibrate-Thrust	(15/23)	2.56*	
Lateral Approach	(16/22)	2.31*	
Horizontal Extend	(17/21)	2.08*	
Following Walk	(18/20)	1.89*	

***$p<.001$
**$p<.01$
*$p<.05$

"Run/Retreat" behaviors, as well as by Contact, Tapping, Wave and Arch, Mutual Approach, and Oblique Extend. Although Intermediate and Subordinate spiders did not differ in the frequency of Run and Retreat, these behaviors differentiated both groups from Dominant spiders. With Dominant opponents, Intermediate spiders were approach-oriented and made Contact preliminary to resolving rank differences. Toward Subordinate opponents, Intermediate spiders exhibited Oblique Extend, the only time a foreleg posture typical of Dominant spiders was exhibited.

Although Q-factor analysis placed Intermediate spiders as polar extremes of Factor I (Subordinance) and Factor II (Dom-

TABLE 6

A comparison of the discriminating ability of behaviors observed during interactions between Dominant (DOM) and Intermediate (INT) Schizocosa crassipes. *[Data modified from Aspey (In Press$_a$) with permission from* Behaviour*]*.

Behavior	(df)	F-value and Probability Level	
		DOM>INT	INT>DOM
Retreat	(2/36)		3.81*
Mutual Approach	(6/32)		2.50*
Prolonged Wave	(7/31)	3.00**	
Contact	(8/30)		2.55*
Tapping	(9/29)		2.47**
Diagonal Approach	(10/28)	2.55**	
Wave and Arch	(11/27)	2.52**	
Run	(12/26)		2.30**
Posterior Approach	(13/25)	2.05*	
Mutual Avoid	(14/24)	1.85*	

**$p<.025$
*$p<.050$

inance), the discriminant analysis considered them a discriminable group. This problem of congruency between different multivariate techniques when applied to the same ethological data has been considered in detail by Aspey and Blankenship (1976c; In Review), and will be considered in the STATISTICAL TALES portion of this paper. Regarding the present data, the two methods for defining/separating subgroups are differentially sensitive to sampling variability inherent within the data. However, no difficulty arises in resolving these differences or in interpretation since the Intermediate spiders of the discriminant analysis systematically corresponded to the polar extremes of Factors I and II. In fact, the use of several analysis techniques is recommended to obtain more complete insight into data organization.

D. Directions for Future Research

With information gained from the various multivariate techniques, hypothesis-seeking for future research can now be more specific and directed. For example, now that Dominant, Intermediate, and Subordinate spiders: (1) are known to exist; (2) can be reliably recognized; and (3) can be opera-

TABLE 7

A comparison of the discriminating ability of behaviors observed during interactions between Intermediate (INT) and Subordinate (SUB) Schizocosa crassipes. *[Data modified from Aspey (In Press$_a$) with permission from* Behaviour*].*

Behavior	(df)	F-value and Probability Level	
		INT>SUB	SUB>INT
Oblique Extend	(5/33)	2.89*	
Mutual Approach	(6/32)	3.01**	
Prolonged Wave	(7/31)		2.62*
Contact	(8/30)	2.56*	
Tapping	(9/29)	2.20**	
Diagonal Approach	(10/28)		1.99*
Wave and Arch	(11/27)	1.97*	

**p<.025
 *p<.050

tionally defined in terms of their behavioral repertoires, hypotheses can be formulated regarding the mechanism(s) of dominance (genetic? developmental?), and the adaptive significance of agonistic display(s) and such distinct dominance classes.

To gain a functional perspective into the agonistic display of S. crassipes, ethologically-relevant environmental parameters were varied (Aspey, In Press$_b$). Using the 20 agonistic behaviors as a "behavioral assay", the influence of different social densities (number of spiders present), spatial densities (amount of space available), and population densities (amount of space available per spider independent of social and spatial density) were assessed using discriminant analysis. Data suggested that the behavioral repertoire serves either to disperse or attract spiders so that a relatively constant inter-individual personal space is maintained. With this information, testable hypotheses can be generated that focus on the advantages such spacing and dominance classes might provide, such as more efficient mating, food utilization, or resource partitioning.

To illustrate one reasonable and testable hypothesis, agonistic display in adult male S. crassipes may maximize the probability of Dominant spiders mating. During a given one-minute observation period in the natural habitat, a male S. crassipes encounters three times as many males as females (Aspey, 1976a). Courtship in S. crassipes is brief, almost a "rape", and contrasts not only with this species' elaborate agonistic display, but also with the remarkably complex court-ship display of other lycosid spiders which frequently lasts several hours (Rovner, 1968). If a male can drive intruding males away from a female without leaving her vicinity, the chances of his mating are enhanced; if the male left the fe-male to chase other males away, rival males are more likely to encounter the female and mate. Thus, given the species' abundance in circumscribed areas and maintenance of a rela-tively constant personal space, both brief courtship and complex agonistic signals would be favored (selected for) that served to drive rival males away while simultaneously allowing a displaying male to remain in the female's vicinity.

Horizontal Extend appears to be one such agonistic sig-nal in that males Retreat or Run 72.6% of the time from a con-specific exhibiting this posture, thereby allowing the dis-playing male to quickly court, mount, and copulate with a nearby female. In conjunction with field data (Aspey, 1976a), a reasonably complete framework of S. crassipes' agonistic communication system emerges, and field and laboratory exper-iments can now be designed to test whether dominance rank con-fers differential mating advantages.

II. ... & SNAILS

Recently, neuroscientists have recognized and exploited the unique research potential of sea hares, vestigial-shelled marine gastropod molluscs of the genus Aplysia. These hermaphroditic snails are important as model systems in neuro-biological and neurochemical research due to their giant, identifiable, pigmented, and topographically organized ganglionic neurons (Frazier, Kandel, Kupfermann, Waziri, and Coggeshall, 1967; Blankenship and Coggeshall, 1976). However, in contrast to the wealth of neurobiological data on Aplysia (see Kandel, 1976) obtained from the relatively constant neuronal geometry and circuitry of the abdominal ganglion, little information is available on Aplysia ethology. Of the more recent behavioral studies (Carew and Kupfermann, 1974; Kupfermann, 1974; Kupfermann and Carew, 1974; Audesirk, 1975; Hamilton and Ambrose, 1975; Aspey and Blankenship, 1976a,b; Aspey, Cobbs, and Blankenship, 1977), one important trend has surfaced; Aplysia behavior is extraordinarily variable under both field and laboratory conditions. Consequently, analytical methods for dealing with this variability must be used so that useful information can be gained from subsequent manipulation of experimental variables.

We have been studying the behavioral biology of Aplysia brasiliana Rang, a sea hare commonly found along the Gulf

Coast that swims by rhythmic undulations of the parapodia
(Hamilton and Ambrose, 1975; Aspey et al.,1977) and burrows in
the sandy substratum (Aspey and Blankenship, 1976a,b). Although little intra- or inter-individual variation exists with
regard to the sequence of movements involved in burrowing,
tremendous individual variation was evident in how different
animals execute burrowing movements. Typically, Aplysia
burrow by pushing the oral tentacles (Figure 5) into the sand

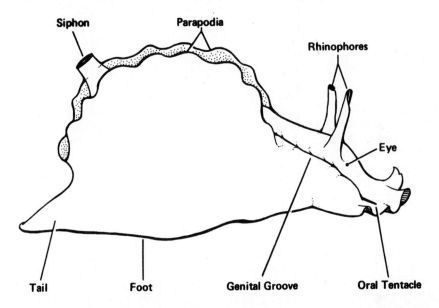

Fig. 5. Diagrammatic representation of pertinent morpho-
logical features of Aplysia brasiliana Rang. [Reprinted from
Aspey and Blankenship (1976a) with permission from Behavioral
Biology].

and then lifting the head upward in a series of "shoveling"
movements that scoop sand forward and laterally around the
body. Once the head is covered, the animal begins "balloon-
ing" and "heaving", a series of forceful pushes into the sand
that alternately "inflate" and "deflate" the body until the
entire animal is covered by sand except the distal tips of the
rhinophores and the dorsal-most fringes of the parapodia.
      Preliminary hypotheses concerning the functional signifi-
cance of burrowing were unsupported: (1) defense from predators: burrowing was sufficiently slow (range 3.17-33.23 min)
so as to be ineffective escape from predators, of which none
are known for Aplysia in the size range studied (15-247 gr);
and (2) response to environmental changes: burrowing was
seemingly unaffected by systematic alterations of water tem-

TABLE 8

Burrowing parameters of Aplysia brasiliana and scores ranges for the different units of measurement or rating scales used to quantify each parameter. Operational definitions of each burrowing parameter given in Aspey and Blankenship (1976a).

Burrowing Parameters	Quantification Units and Range
Weight	15-247 g
Latency	1.02-28.52 min
Burrowing Time	3.17-33.23 min
Time Burrowed	0.04-15.33 days
Inking	1-3
Condition	1-3
Vigor	1-4
Emergence	0-6
Coverage	1-6
Direction	1-6

perature, salinity, and/or light/dark cycles within biologically meaningful ranges. To gain insight into "how and why" Aplysia burrow, we identified and operationally defined 10 parameters of burrowing animals in terms of (1) direct measurements of weight and time duration, and (2) intensity ratings of behaviors associated with burrowing on variable interval scales according to specified criteria. These burrowing parameters and their quantification are summarized in Table 8 and operationally defined in Aspey and Blankenship (1976a). Since one of our goals was to identify relationships among naturally-selected heterogeneous subjects when the underlying biological basis of individual variation was unknown, we used a variety of multivariate analyses to determine inherent structure within the data.

Our purpose in the SNAILS portion of this chapter is to demonstrate the usefulness of Q-type orthogonal powered-vector factor analysis and linear typal analysis for identifying biologically interpretable homogeneous subgroups of Aplysia. From these subject-related subgroups, hypotheses were generated and tested (Aspey and Blankenship, 1976b) that not only provided insight into the functional significance of burrowing, but also suggested the existence of pheromone(s) controlling burrowing, sexual behavior, and/or aggregation.

A. Determining Structure Among the Snails

Q-Factor Analysis and Linear Typal Analysis

The data were: (1) collected from 32 A. brasiliana Rang during 76 separate test observations of burrowing, each lasting 6-7 hr, over a 3-mo period; and (2) cast into a matrix of 10 rows (burrowing parameters) x 32 columns (snails), similar to the Q-factor analysis for the SPIDERS. In order to identify natural groupings among the snails (i.e., subject-related factors) from data quantified in arbitrary and/or disparate units of measurements, we used Q-type orthogonal powered-vector factor analysis with a cluster-oriented solution (hereafter termed Q-factor analysis) and linear typal analysis (Overall and Klett, 1972).

In contrast to principal-axis factor analysis discussed earlier, the powered-vector method (Overall and Porterfield, 1962) tends to position factors meaningfully without rotation, placing less emphasis on parsimony (i.e., maximum variance accounted for by a few factors) and more emphasis on biological and/or psychological relevancy of the factors. However, the powered-vector method does yield an orthogonal "cluster-oriented" solution (Overall, 1968) in which each factor represents a distinct homogeneous subset of variables (or subjects when using Q-factor analysis). Linear typal analysis is an empirical classification method mechanically similar to factor analysis except for minor scaling differences. Theoretically, linear typal analysis assumes that a relatively few basic hypothetical "pure-types" underlie any heterogeneous group, and aims to determine the number and nature of these pure-types and their relationship to the subject (Overall and Klett, 1972).

Data Transformations

Since the burrowing parameters were quantified in arbitrary and/or disparate units of measurement (Weight in grams; Latency in minutes; Time Burrowed in days; Inking on a scale 1-3; Emergence on a scale 0-6), the data were transformed to eliminate such "statistical variance" (variance due to different units of measurement). Furthermore, since different data transformations can influence the number of subgroups that emerge, all scores were transformed in the same way so as to facilitate comparison between Q-factor analysis and linear typal analysis.

Some investigators prefer raw scores for generating vector-product matrices for multivariate analyses (Nunnally, 1962; Tucker, 1968), but since ethologists are frequently in-

terested in studying individual differences, transformed data
are more practical (Gollob, 1968). Common data transformations include: (1) ranking; (2) standardizing; (3) normalizing; and (4) origin-correcting. Ranking is preferred for
scaled data with uniform scale ranges five or more units
(i.e., all variables scaled 1-6). Standardizing to z-scores
across individuals allows meaningful comparisons among variously scaled data by eliminating such disparate units of
measurement. Normalizing transformed scores to unity in the
diagonal further helps account for individual differences, and
not just statistical variance, due to diverse units of
measurement along the variables. Finally, origin-correcting
aids in demonstrating meaningful pattern differences when row
and column effects due to arbitrary zero-points are removed.
Typically, origin-correction avoids cases where (1) there results one less than the number of homogeneous subgroups
actually represented in the data, or (2) only one large general factor emerges. Therefore, the data on burrowing were
(1) transformed to z-scores across subjects, (2) corrected to
zero-mean across subjects with a constant added ($\underline{k} = 0.5623$)
to provide an origin that accounted maximally for individual
differences (Overall and Klett, 1972), and (3) the resulting
transformed scores were then normalized to unity in the
diagonal. Overall and Klett (1972) discuss the rationale and
mechanics of several data transformations, most of which are
user-selected options on available computer statistical
packages.

Extracted Homogeneous Subgroups

Q-factor analysis applied to the transformed data matrix
extracted three factors with eigenvalues ≥ 1.0 that accounted
for 80.2% of the variance. Each factor represents a distinct
homogeneous subset of snails and all snails within each factor
correlate highly with the dimension represented by that factor
(Figure 6). Unlike Tables 2 and 3 where only factor loadings
≥ 0.45 are listed, Figure 6 lists all factor loadings to
emphasize that each factor represents a different source of
individual variation, as evidenced by each snail having a high
factor loading on only one factor and near-zero loadings on
the remaining factors. Figure 7 illustrated a three-dimensional projection of the snails by plotting the factor loadings as coordinates on X, Y, and Z axes, which correspond to
Factors I, II, and III. The three subject-related subgroups
show distinct within-group clustering as well as between-group
separation.

Linear typal analysis assigns subjects to different extracted pure-types and does not provide a factor loading
to indicate the extent a given subject correlates

Subjects	Original Pure-Type	Factors I	II	III
26	1	0.99	0.08	-0.03
29	1	0.97	0.05	0.06
32	1	0.96	0.06	-0.06
27	1	0.91	0.05	0.17
31	1	0.91	-0.04	0.01
25	1	0.90	0.08	0.14
28	1	0.90	-0.06	0.17
24	1	0.84	0.25	-0.26
30	1	0.76	0.31	-0.13
23	3	-0.06	0.27	0.87
19	3	-0.31	0.56	0.76
21	3	0.33	-0.04	0.70
6	3	-0.34	0.60	0.68
13	3	-0.07	0.31	0.64
20	3	-0.05	0.32	0.62
9	3	-0.26	0.33	0.59
22	3	0.03	0.46	0.59
16	2	-0.18	0.91	-0.20
18	2	-0.26	0.90	-0.08
17	2	-0.15	0.87	-0.07
8	2	-0.46	0.84	0.04
12	2	-0.42	0.80	0.01
4	2	-0.42	0.79	0.02
7	2	-0.41	0.78	0.03
1	2	-0.35	0.71	0.14
10	2	-0.41	0.65	0.30
2	4	-0.34	0.64 (0.68)	0.01
15	2	-0.45	0.64	0.34
3	2	-0.29	0.62	0.12
11	2	-0.25	0.57	0.28
14	4	-0.36	0.51 (0.68)	0.28
5	4	-0.21	0.46 (0.75)	0.07

Fig. 6. A comparison of the classification of 32 Aplysia brasiliana according to 10 burrowing parameters by Q-type orthogonal powered-vector factor analysis with a cluster-oriented solution and linear typal analysis. Within the three independent factors extracted by Q-factor analysis, the subjects comprising a given factor are ordered according to their factor loadings on that factor. These results are compared with the classification of the same subjects into four pure-types by linear typal analysis. Although agreement between these two multivariate techniques is noteworthy, pure-types 2 and 4 were not orthogonal and were consolidated for interpretation purposes. Factor loadings on a fourth nonorthogonal factor (Factor IV) are given in parentheses for those subjects (#2, 5, and 14) comprising pure-type 4. These snails simply seem to represent one extreme of pure-type 2, which also corresponds to Factor II.

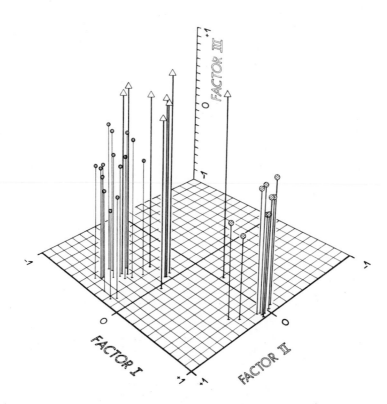

Fig. 7. Factor loadings from Figure 6 projected onto coordinate axes corresponding to the three factors extracted by Q-factor analysis. The origin falls in the center of the three-dimensional space. Factor I (large striped circles) represents "Inefficient Burrowers"; Factor II (small stippled circles) "Efficient Burrowers"; and Factor III (triangles) "Intermediate Burrowers". [This figure modified from Aspey and Blankenship (1976a) and reprinted with permission from Behavioral Biology*].*

with the dimension represented by the pure-type. Referring to Figure 6, linear typal analysis placed 29 of 32 snails (91%) into three pure-types that corresponded remarkably well to the three factors, while the remaining three subjects (#2, 5, and 14) formed pure-type 4. As factor analysis defines statistically uncorrelated factors that represent reasonably distinct dimensions of variation, pure-types extracted by linear typal analysis are considered independent if the matrix of "cosines" (intercorrelations among the pure-types; Table 9) is an identity matrix (i.e., unity in the principal diagonal and low or near-zero correlations in the remaining cells). Since

TABLE 9

Cosine matrix (intercorrelations) of four pure-types extracted by linear typal analysis that represent four subgroups of burrowing Aplysia brasiliana. Correlations (r_s) not in italics represent orthogonal pure-types. Therefore, pure-types 2 and 4 are not statistically independent and were consolidated for interpretation purposes.

Pure-Types	1	2	3	4
1	*1.000*			
2	-0.305	*1.000*		
3	-0.038	0.486	*1.000*	
4	-0.275	*0.705*	0.463	*1.000*

pure-types 2 and 4 were highly correlated (r_s = 0.705), and therefore not statistically independent, they were consolidated for interpretation purposes (Overall and Klett, 1972). Figure 6 shows that pure-type 4 seems to represent one extreme of pure-type 2 (or Factor II).

Interpretation of the Extracted Subgroups

The three subgroups were interpreted on the basis of the burrowing characteristics unifying the snails within each group. Factor I accounted for 36.7% of the variance and the nine snails comprising this factor had either the highest or lowest scores on nine of the 10 burrowing parameters (highest: Weight, Burrowing Time, Latency, Time Burrowed, Direction; lowest: Coverage, Inking, Vigor, Condition). These snails were the heaviest, slowest, and least vigorous animals that burrowed incompletely, responded minimally to disturbances during burrowing, and remained burrowed for at least a week. On this basis, Factor I was interpreted as "Inefficient Burrowers", perhaps representing animals that were old and/or in poor health. This interpretation was strengthened by the observation that all of these animals were dead 10 days after the last observed bout of spontaneous burrowing.

Factor II accounted for 24.6% of the variance and the 15 snails comprising this factor were also characterized by relatively extreme scores on seven of the 10 burrowing parameters, but in the opposite direction of Factor I (highest: Emergence, Coverage, Inking, Vigor; lowest: Weight, Burrowing Time, Latency). These snails were the smallest, fastest, and most vigorous animals that burrowed completely, responded

maximally to disturbances during burrowing, and remained
burrowed for only a few days. On this basis, Factor II was
interpreted as "Efficient Burrowers", perhaps representing
young, reproductively active and/or healthy Aplysia. This interpretation was strengthened by the observation that all of
these snails were alive and actively swimming and/or copulating 10 days after the last observed bout of spontaneous
burrowing.

        Finally, Factor III accounted for 18.9% of the variance
and consisted of eight Aplysia which exhibited relatively low
scores on six of the 10 burrowing parameters (Time Burrowed,
Direction, Emergence, Coverage, Inking, Vigor), and medium
scores on the remaining parameters (Weight, Burrowing Time,
Latency, Condition). These snails were active, burrowed
rhythmically and rapidly, but remained burrowed for less than
a day. When compared to subjects in Factors I and II, Factor
III seemed to represent snails sharing more characteristics
with Efficient Burrowers than with Inefficient Burrowers, as
spatially illustrated in Figure 7. Factor III was interpreted
as "Intermediate Burrowers", perhaps representing snails in
transition from being Efficient Burrowers to becoming Inefficient Burrowers. This interpretation is plausible since
physiological and/or maturational transitional states are
common in molluscs, such as sex changes in the slipper-limpet
Crepidula (Coe, 1936). More complete and detailed characterization of these burrowing types are presented in Aspey and
Blankenship (1976a).

B.   Directions for Future Research

        Burrowing in A. brasiliana was more complex than origin-
ally anticipated, and one advantage of the statistical analyses employed was the differentiation of naturally-sampled,
heterogeneous subjects into homogeneous subgroups on the basis
of multiple behavioral parameters. The identified subgroups
now provided a quantitatively reliable method for selecting
suitable subjects for testing specific hypotheses. For example, we hypothesized that Efficient Burrowers were young and/
or healthy animals for whom burrowing may serve as a preparatory state for subsequent reproductive activity since these
animals aggregate, copulate, and/or lay eggs following emergence from burrowing. Conversely, Inefficient Burrowers were
probably old and/or unhealthy animals for whom burrowing may
represent an energy-conserving response to deteriorating
health or lowered tolerance to unfavorable environmental conditions since all these animals died within 10 days following
burrowing with no involvement in sexual behavior. Intermediate Burrowers seem to represent animals in transition from

being Efficient Burrowers to becoming Inefficient Burrowers.

We selected 12 Efficient Burrowers to examine those conditions that initiate, maintain, and terminate burrowing (Aspey and Blankenship, 1976b). With regard to the initiating stimulus, most animals appeared to be intrinsic burrowers, suggesting that burrowing was a maturational phenomenon. However, on several occasions when an Efficient Burrower was placed with a larger (heavier than 250 g), nonburrowed conspecific in a small aquarium, the larger animal made burrowing "attempts" within 30 min to the extent that its head was covered with sand. This observation was unusual since burrowing in such large animals has not been seen previously. Upon returning these animals to the holding tank we noticed that the Efficient Burrower reburrowed, while the induced burrower (i.e., Nonburrower) did not.

Our results (Aspey and Blankenship, 1976b) indicated that although a pheromone alone is sufficient to initiate burrowing in Nonburrowers, it is not sufficient to maintain burrowing since Nonburrowers induced to burrow did not persist in reburrowing following removal from the burrow. In contrast, over 80% of the Efficient Burrowers not only reburrowed for 10 removals, but also initiated burrowing faster and faster with each removal. Additional evidence further suggested that some "internal state" is necessary for the maintenance of burrowing once triggered. Therefore, burrowing was at least a two-process phenomenon involving a physiological "internal readiness" to burrow as well as an external pheromone component. Whether a pheromone is biologically "necessary" to initiate burrowing in animals that do not burrow spontaneously remains unclear. However, such a pheromone represents one mechanism for synchronizing burrowing at the population level for those animals internally prepared to burrow. Given the evidence that Efficient Burrowers engage in reproductive activities (i.e., egg-laying, aggregating, copulating) upon spontaneous emergence from burrowing suggests one advantage for a mechanism that insures burrowing in close spatial proximity among conspecifics.

Although the nature of the putative pheromone and suspected internal state is unknown, subsequent experiments (Aspey and Blankenship, in preparation) have shown that a conspecific in the process of egg-laying is the source for terminating burrowing and/or releasing sexual behavior. The hypothesis that Efficient Burrowers represent young and/or healthy Aplysia for whom burrowing is a preparatory period for subsequent bouts of reproductive activity was strengthened by observations that Efficient Burrowers emerged from the sand and copulated each time an egg-laying conspecific was present. This suggested that a different pheromone may override the internal state of the animal and trigger sexual behavior with

the subsequent termination of burrowing. Similarly, Davis
and Mpitsos (1971) have suggested that a chemosensory cue may
incite sexual behavior in the related marine gastropod
Pleurobranchaea californica. Furthermore, Hamilton and Ambrose's (1975) observation that abundant numbers of A.
brasiliana suddenly appear in the natural habitat suggest some
form of behavior serving to bring conspecifics together. Our
additional studies (Aspey and Blankenship, in preparation)
clearly indicate that an egg-laying conspecific produces/
releases a pheromone(s) that facilitates aggregation and increases the probability of copulation and/or egg-laying. Pre-
sumably, this phenomenon could serve as one mechanism for
bringing conspecifics together and maintaining breeding
aggregations in the field.

Thus, the use of multivariate statistical techniques have
provided a strong quantitative foundation from which information was gained, both directly and indirectly related to
burrowing. Furthermore, meaningful hypotheses were generated,
some experimentally tested, and insight gained into mechanisms
and the adaptive significance of a variety of biologically
significant behaviors in Aplysia.

III. ... & STATISTICAL TALES

As in other analytical efforts, a diversity of approach
is desirable to produce the most complete understanding of a
system's dynamics. In the SPIDERS portion, for example, Q-
factor analysis identified two subgroups while the DI identified three; discriminant analysis was then used to verify
the distinctness of the three groups. In the SNAILS portion,
Q-factor analysis identified three subgroups while linear
typal analysis identified four, however, pure-type 4 was
nonorthogonal with pure-type 2 and simply seemed to represent
one extreme of pure-type 2. One aspect of animal behavior
research that arouses considerable uncertainty is knowing
which quantitative analysis to use. Faced with a variety of
statistically legitimate analyses, the researcher must select
the one most useful in terms of biologically/psychologically
meaningful interpretability. This statement re-emphasizes
two points made previously: (1) fit the method of analysis to
the animal and the problem under consideration (see also
Bekoff, this volume); and (2) use multivariate analysis as a
diagnostic tool for uncovering inherent structure within data,
and for identifying the relevancy of certain variables.
Therefore, it is perfectly appropriate to use a variety of
quantitative methods in order to assess the usefulness of any
one or two analyses in terms of gaining knowledge or insight
for answering specific questions or for directing future

research.

To this extent, the STATISTICAL TALES portion of this chapter compares the congruency of five different multivariate analyses in terms of the number of extracted subgroups and their biological interpretability when applied to the same ethological data. A preliminary report of these findings is abstracted in Aspey and Blankenship (1976c), additional details are given in an expanded version by Aspey and Blankenship (In Review), and more theoretical and mechanical statistical considerations are presented in Overall and Klett (1972).

A. Comparative Ethometrics

The analyses discussed here are empirical classification typologies for deriving homogeneous subgroups from measurements on multiple variables without regard for any previously existing classification scheme. The same transformed data matrix as used in the SNAILS portion will be analyzed by the following multivariate techniques: (1) multidimensional scaling; (2) principal-components analysis; (3) Q-factor analysis; (4) linear typal analysis; and (5) cluster analysis. Q-factor analysis and linear typal analysis were also discussed in the SNAILS portion and those results are used as a frame of reference for comparing/contrasting the variously extracted subgroups from the remaining three analyses.

Multidimensional Scaling

Multidimensional scaling represents similarities and differences among subjects or concepts in terms of distances along coordinate axes. Data applicable to multidimensional scaling range from subjectively judged similarities and differences (e.g., "greater than" or "less than" comparisons) to calculations of similarity indices from measurement profiles (Overall and Free, 1972; Shepard, Romney, and Nerlove, 1972). The data were analyzed using the multidimensional scaling program developed by Overall and Free (1972) for relating clinical symptom-indication data to models of psychotherapeutic drug action established on pre-clinical test animal profiles. This program is flexible and can be used when neither objects nor variables are segregated into prior groups, as in the present study. Overall and Free's (1972) multidimensional scaling is essentially a principal-components analysis that displays relationships among all subjects (or variables) on the basis of a preliminary analysis of one subset. Consequently, the following discussion of principal-components analysis also applies to multidimensional scaling. Information on multi-

dimensional scaling from an ethological perspective is
available in Morgan et al. (1975).

Interestingly enough, multidimensional scaling ordered
the subjects along a single continuum expressed as vector
scores from -1 to +1 and interpreted as a dimension of
"burrowing efficiency". Although multidimensional scaling indicated only one group of subjects, a general correspondence
existed between the subjects' ordering by multidimensional
scaling and the three Q-factor analysis subgroups (hereafter
termed FA 1, FA II, and FA III). For example, subjects near
the -1 end of the continuum represented subjects comprising
FA 1 (Inefficient Burrowers, snails #29, 28, 26, 32, 25, 31,
27, 30). Consistent with expected results, _Aplysia_ #29 and
#26, with the lowest multidimensional scaling scores
(-1.000, rank 32 of 32; and -0.766, rank 30 of 32, respectively) had the highest factor loadings on FA 1 (0.97 and
0.99, respectively), indicating that these animals were the
prototypical Inefficient Burrowers.

Subjects near the +1 end of the continuum represented
animals comprising FA II (Efficient Burrowers, subjects #1, 8,
11, 2, 4, 7, 14, 10, 15, 18, 16, 3, 5, 17, 12). However,
in spite of relative correspondence between multidimensional
scaling subject-ordered and factor loadings on FA I, inconsis-
tencies were apparent for FA II. For example, the rating of
Aplysia #1 with the highest positive multidimensional scaling
score (0.860, rank 1 of 32) was puzzling since this snail
placed near the middle of FA II (factor loading = 0.71) and
did not have the highest factor loading, as expected. Conversely, _Aplysia_ #16, with the highest FA II factor loading
(0.91), should have had the highest multidimensional scaling
score instead of 0.230 (rank 13 of 32). Additionally, subject
#24 from FA I was placed among FA II subjects by multidimen-
sional scaling. Finally, subjects near the middle of the
multidimensional continuum corresponded to FA III (Intermediate Burrowers, subjects #21, 20, 22, 23, 13, 9, 19, 6).
As expected, _Aplysia_ #23, with the highest factor loading
(0.87) on FA III, had a multidimensional scaling score of
-0.10 (rank 21 of 32) placing it near the middle of the continuum indicating that this animal was the prototypical Intermediate Burrower.

Principal-Components Analysis

Principal-components analysis describes differences between individuals in a heterogeneous sample in terms of a
relatively few composite variables (i.e., components). Therefore, principal-components analysis reduces correlated variables to a smaller set of statistically independent linear
combinations which characterize individual differences (Over-

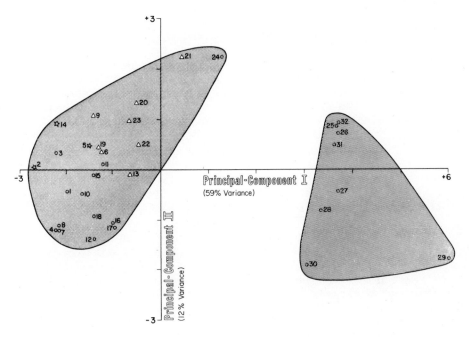

Fig. 8. Principal-components analysis of 32 burrowing Aplysia brasiliana showing two extracted subgroups. When compared to the Q-factor analysis, the first principal component (striped area) accounted for 12% of the variance and generally corresponded to the first half of FA II and FA III (triangles). Subjects represented by stars corresponded to pure-type 4 extracted by linear typal analysis. [This figure reprinted from Aspey and Blankenship (In Press$_a$) with permission from Animal Behaviour].

all and Klett, 1972). The first principal component is that weighted combination which accounts for a maximum amount of total variation (i.e., individual difference) in the complete set of original variables. The second principal component is that weighted combination of the original multiple variables, which of all possible weighted combinations uncorrelated with the first, accounts for the greatest possible proportion of the remaining variation. Principal-components analysis makes no assumptions about the distribution of the variables. However, if several variables have disproportionately large variance, the principal component solution will be "pulled" in the direction of accounting for these variances. Thus, for principal-components analysis to maximally account for individual differences, and not just statistical variance, the data matrix should be transformed with unity in the diagonal, as we have done. Applications of principal-components analysis to animal behavior include studies by Brothers and Michener (1974), Halliday (1976), and Huntingford (1976a,b).

Principal-components analysis identified two subgroups from the 32 burrowing A. brasiliana that accounted for 71% of the variance (Figure 8). The first principal component (striped area) accounted for 59% of the variance and generally corresponded to subjects comprising FA I (striped circles, except subject #24) and the second half of FA II. The second principal component (stippled area) accounted for 12% of the variance and corresponded to subjects comprising the first half of FA II and FA III. Although principal-components analysis failed to differentiate three subgroups, as Q-factor analysis did, Figure 8 shows a relative grouping of FA III Aplysia (triangles) above the abscissa while the FA II animals (stippled circles) group below the abscissa. Essentially, the two principal components can also be interpreted as dimensions of "burrowing efficiency" with the first principal component representing Inefficient Burrowers (FA I and the second half of FA II), and the second principal component representing Efficient Burrowers (the first half of FA II and FA III).

Cluster Analysis

Cluster analysis results in several mutually exclusive subgroups (clusters) within which individuals are relatively similar and between which individuals are relatively different. The most generally useful index of multivariate similarity is the simple distance-function ("d" measures of dissimilarity), calculated as the sum of squares of differences between corresponding scores in two multivariate profiles. For simple distance-function calculations, each original measurement variable is associated with a distinct orthogonal axis and differences between two individuals' scores are conceived of as differences in projections on the orthogonal axes. Cluster analysis methods often make fewer assumptions than other multivariate techniques, are easier to understand, and allow a diversity of data representations, including numerical characterization using d values, or visual representations using dendograms, maximum spanning trees, and three-dimensional geometric models. Morgan et al. (1975) review cluster analysis techniques as applied to ethological data, while Schwartz et al. (1976) use cluster analysis as a tool in neuroethology.

Cluster analysis identified five homogeneous clusters using simple distance functions (d) (Table 10). The principal diagonal represents within-cluster distances (italics), while the off-diagonals represent between-cluster distances. Table 10 shows homogeneous within-group clustering for all five clusters since the within-cluster distances averaged less than

TABLE 10

Simple within-cluster (italics) and between-cluster distances for five clusters extracted by cluster analysis that represent subgroups from 32 Aplysia brasiliana.

Clusters	1	2	3	4	5
1	*0.55*				
2	0.85	*0.57*			
3	0.89	0.92	*0.58*		
4	0.87	0.90	0.96	*0.54*	
5	0.63	0.84	0.77	0.69	*0.26*

one-half the between-cluster distances for every cluster pair. Figure 9 illustrates how within- and between-cluster distances are used to position and relate the clusters in a reduced geometric space. A satisifying cluster configuration is achieved when the mean between-cluster distances exceed the average within-cluster distances by 3:1 or 4:1 (Overall and Free, 1972). For these data, the mean within-cluster distance is 0.26, and only the between-cluster distances among the first four clusters met this criterion. Although cluster 5 (N = 2, subjects #14 and #9) showed acceptable within-group homogeneity, this subgroup did not relate well geometrically to the other clusters (i.e., distances between cluster 5 and the other four clusters were not greater than three or four times the mean within-cluster distance).

Although clusters 1, 2, and 4 generally correspond to FA II, FA I, and FA III, respectively, several classification difficulties exist. For example, animals comprising cluster 3 (N = 5, subjects #30, 18, 17, 3, 13) and cluster 5 (N = 2, subjects #14 and #9) are dispersed unsystematically among the three Q-factor analysis subgroups (FA I has one subject from cluster 3; FA II has three animals from cluster 3 and one subject from cluster 5; and FA III has one subject each from clusters 3 and 5). Furthermore, the cluster analysis failed to classify subjects #12 and #24 within any of the five clusters.

B. Analysis Sensitivity and Interpretation Usefulness

Given the divergence in the number of subgroups extracted by different multivariate analyses (from one to five) when applied to the same transformed data matrix, confusion among

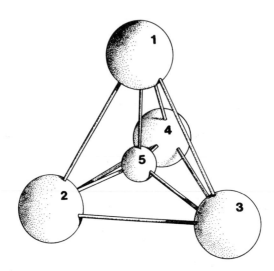

Fig. 9.   A three-dimensional representation of Table 10 showing five clusters extracted by cluster analysis from 32 burrowing Aplysia brasiliana. Each cluster's diameter corresponds to within-group distances (d), while distances between clusters correspond to between-group distances. Subjects comprising clusters 1, 2, and 4 generally corresponded to FA II, FA I, and FA III, respectively. Subjects comprising cluster 3 were dispersed among the three Q-factor analysis groupings. Of the two animals comprising cluster 5, one each was from FA II and FA III. Due to the two-dimensional artistic perspective of three dimensions, some distortion of the actual within- and between-cluster distances is present. [This figure reprinted from Aspey and Blankenship (In Press) with permission from Animal Behaviour].

students and researchers in deciding which quantitative method to use is not surprising. However, most differences in subject placement within the variety of extracted subgroups could be resolved with relatively consistent interpretations of "burrowing efficiency". Such diversity in the number of extracted subgroups is, nevertheless, a strong argument in favor of using several multivariate analyses in concert so as to provide a more flexible base for complete data interpretations. Furthermore, all analyses reinforced one another while each added a subtly different perspective into suggesting possible mechanisms underlying behavior.

To illustrate these points, if only the multidimensional scaling results were available, conceptualizing burrowing along an efficiency continuum would probably result. However, gaining insight into the significance of the variability in any systematic way would probably be more difficult to achieve. With only the two principal components available, a likely interpretation might involve two genetically discrete groups. Although such an interpretation is plausable given the wide genetic variability of molluscs, subsequent longitudinal data did not support this interpretation. The three Q-factor analysis groups indicated not only relatively distinct burrowing "types", but also suggested that burrowing was a dynamic process. Although using the multidimensional scaling and principal-components analysis in conjunction with one another would have likely led to a similar conclusion, the Q-factor analysis subgroupings permitted this conclusion more quickly and parsimoniously. Linear typal analysis and Q-factor analysis yielded nearly identical results, with pure-type 4 representing one extreme of FA II. Finally, testimony into the wide variability inherent within the sample was evidenced by the single multidimensional scaling group and uninterpretability of two of the five cluster analysis subgroups.

In this research example, several advantages were gained by using multivariate analyses as diagonistic tools for data organization and hypothesis-seeking: (1) burrowing appeared as a dynamic process which, nonetheless, allowed the identification of distinct subgroups; (2) testable hypotheses were generated concerning the underlying mechanisms of burrowing for these various subgroups; (3) insight was provided into the functional significance of burrowing for the different subgroups; and (4) the opportunity was afforded to perform experiments using the now-identified homogeneous subgroups. This last advantage was specifically demonstrated by Aspey and Blankenship's (1976a) discovery of a pheromone synchronizing burrowing in Efficient Burrowers that induced burrowing in conspecifics having near-zero probabilities of spontaneous burrowing. Furthermore, burrowing seemed to be a two-process phenomonen consisting of: (1) a pheromone capable of triggering the behavior; and (2) an internal state that maintains the behavior. These findings suggested how the adaptive significance of burrowing might differ for Efficient and Inefficient Burrowers (see Aspey and Blankenship, 1976a,b), while Intermediate Burrowers provided the basis for hypothesizing a transitional, internal maturational- and/or health-linked state by which Efficient Burrowers become Inefficient Burrowers.

The use of a variety of different analyses also revealed consistently troublesome subjects to classify. For example, subjects #12 and #24 were: (1) not included in any of the five cluster analysis groups; (2) inconsistently classified by principal-components analysis relative to the other analyses; and (3) placed near the middle of the continuum by multidimensional scaling (see Aspey and Blankenship, In Rev.). Subject #24's scores were more similar to those of Efficient Burrowers on Weight, Latency, Burrowing Time, and Time Burrowed, but more similar to those of Inefficient Burrowers on the other burrowing parameters. Subject #12's scores were representative of Efficient Burrowers (except for Weight and Latency), and the reason for difficulty in classifying this subject is less obvious.

The five empirical classification typologies discussed here unquestionably differ in their sensitivity to sampling variability within the data. To this extent, guidelines can be suggested for selecting an appropriate analysis. Specifically, linear typal analysis and cluster analysis appear more sensitive to sampling variability relative to the other analyses and tend to "split" subjects into additional groups. Such analyses might be most useful with precise data measurement and minimal sampling variance. Overall and Klett (1972) have also suggested that cluster analysis methods seem especially sensitive to sampling variability because the starting point for each cluster depends on only two or three profiles out of the total sample. Conversely, multidimensional scaling, principal-components analysis, and Q-factor analysis take into account, to varying degrees, relationships among all subjects in the sample. As such, multidimensional scaling and principal-components analysis seem less sensitive to wide variability and tend to "lump" subjects relative to other analyses. For example, the first two principal components accounted for 71.0% of the variance while the first two factors accounted for only 63.3% of the variance (the third factor accounting for an additional 18.9%). Therefore, multidimensional scaling and principal-components analysis may be more appropriate or experimentally useful for qualitatively scaled data and/or when great heterogeneity of variance exists.

In conclusion, when used in proper perspective, multivariate analyses are potentially powerful research tools: (1) for uncovering homogeneous subgroups from naturally-selected heterogeneous samples, and (2) for hypothesis-seeking. Additionally, in selecting a meaningful analysis, the researcher in animal behavior should be aware of differences in sampling variability among analyses, and select those methods either minimally or maximally sensitive to relationships between pairs of individuals: (1) according to the type of data, (2) appropriate for the problem under consideration, and (3) yielding a reasonable biological/psychological interpretation.

ACKNOWLEDGEMENTS

The SPIDERS portion of this paper represents research by

W.P.A. submitted to Ohio University in 1974 for the Ph.D. degree in Zoology under the direction of Professor Jerome S. Rovner. This research was supported by Hiram Roy Wilson Research Fellowships in Zoology to W.P.A. and grant GB 35369 from the National Science Foundation to J.S. Rovner. The SNAILS & STATISTICAL TALES portions of this chapter were conducted under a postdoctoral fellowship to W.P.A. from The Marine Biomedical Institute, The University of Texas Medical Branch at Galveston, under the sponsorship of J.E.B. This research and the preparation of this manuscript were supported by NIH Grant NS 11255, a grant from the Moody Foundation of Galveston, Texas, and NIH award NS 70613 to J.E.B. It is our sincere pleasure to thank the following colleagues for their generous advice, helpfulness, and expertise during various phases of this research: Drs. Jerome S. Rovner and Gerald Svendsen, Ohio University; Dr. Patrick Colgan, Queen's University, Canada; Dr. Brian A. Hazlett, University of Michigan; Dr. Keith Nelson, Bodega Bay, California; Dr. Harold M. Pinsker, The Marine Biomedical Institute; and Professor John E. Overall, Psychometric Laboratory, the University of Texas Medical Branch at Galveston, for computer facilities and helpful guidance with the statistical analyses.

REFERENCES

Aspey, W.P. 1976a. Behavioral ecology of the "edge effect" in Schizocosa crassipes (Araneae: Lycosidae). Psyche 83: 42-50.

_____. 1976b. Response strategies of adult male Schizocosa crassipes (Araneae: Lycosidae) during agonistic interactions. Psyche 83: 94-105.

_____. In Press_a. Wolf spider sociobiology: I. Agonistic display and dominance-subordinance relations in adult male Schizocosa crassipes. Behaviour.

_____. In Press_b. Wolf-spider sociobiology: II. Density parameters influencing agonistic behavior in Schizocosa crassipes. Behaviour.

_____. In Press_c. Fiddler crab behavioral ecology: Burrow density in Uca pugnax and Uca pugilator. Crustaceana.

_____ and J.E. Blankenship. 1976a. Aplysia behavioral biology: I. A multivariate analysis of burrowing in A. brasiliana. Behav. Biol. 17: 279-299.

_____ and _____. 1976b. Aplysia behavioral biology: II. Induced burrowing in swimming A. brasiliana by burrowed conspecifics. Behav. Biol. 17: 301-312.

_____ and _____. 1976c. Burrowing in Aplysia: Interpretation and application of different multivariate analyses to the same data. Am. Zool. 16: 245 (Abstr.).

_____ and _____. In Review. Comparative ethometrics: Congruence of different multivariate analyses applied to the
same ethological data.

_____, J.S. Cobbs, and J.E. Blankenship. 1977. Aplysia be-
havioral biology: III. Head-bobbing in relation to food
deprivation in A. brasiliana. Behav. Biol. 19: In Press.

Atchley, W.R. and E.H. Bryant (eds.). 1975a. Multivariate
Statistical Methods: Vol. 1. Among-Groups Covariation.
Halsted Press, New York.

_____ and _____ (eds.). 1975b. Multivariate Statistical
Methods: Vol. 2. Within-Groups Covariation. Halsted
Press, New York.

Audesirk, T.E. 1975. Chemoreception in Aplysia californica.
I. Behavioral localization of distance chemoreception used
in food-finding. Behav. Biol. 15: 45-55.

Barlow, G.W. 1968. Ethological units of behavior. p. 217-232
in D. Ingle (ed.) The Central Nervous System and Fish
Behavior, Univ. of Chicago Press, Chicago.

Bekoff, M., H.L. Hill, and J.B. Mitton. 1975. Behavioral
taxonomy in canids by discriminant function analysis.
Science 190: 1223-1225.

Bishop, Y.M.M, S.E. Fienberg, and P.W. Holland. 1975. Discrete Multivariate Analysis. MIT Press, Cambridge.

Blankenship, J.E. and R.E. Coggeshall. 1976. The abdominal
ganglion of Aplysia brasiliana: A comparative morphological and electrophysiological study, with notes on A.
dactylomela. J. Neurobiol. 7: 383-405.

Bristowe, W.S. 1929. The mating habits of spiders, with
special reference to the problems surrounding sex dimorphism. Proc. Zool. Soc. Lond. 1929: 309-358.

Brothers, D.J. and C.D. Michener. 1974. Interactions in col-
onies of primitively social bees. III. Ethometry of
division of labor in Lasioglossum zephyrum (Hymenoptera:
Halictidae). J. Comp. Physiol. 90: 129-168.

Burgess, J.W. 1976. Social spiders. Sci. Am. 234: 100-107.

Buskirk, R.E. 1975. Aggressive display and orb defense in a
colonial spider Metabus gravidus. Anim. Behav. 23: 560-
567.

Carew, T.J. and I. Kupfermann. 1974. The influences of different natural environments on habituation in Aplysia
californica. Behav. Biol. 12: 339-345.

Cassie, R.M. 1969. Multivariate analysis in ecology. Proc.
N. Z. Ecol. Soc. 16: 53-57.

Cassie, R.M. and A.D. Michael. 1968. Fauna and sediments of
an intertidal mud flat: A multivariate analysis. J. Exp.
Mar. Biol. Ecol. 2: 1-23.

Chase, I.D. 1974. Models of hierarchy formation in animal
societies. Behav. Sci. 19: 374-382.

Coe, W.R. 1936. Sexual phases in Crepidula. J. Exp. Zool.

73: 455-477.

Comrey, A.D. 1973. A First Course in Factor Analysis.
  Academic Press, New York.

Cooley, W.W. and P.R. Lohnes. 1971. Multivariate Data Analysis. John Wiley & Sons, New York.

Crane, J. 1949. Comparative biology of salticid spiders at
  Rancho Grande, Venezuela, Part IV. An analysis of display. Zoologica 34: 159-214.

Davis, W.J. and G.J. Mpitsos. 1971. Behavioral choice and
  habitation in the marine mollusc Pleurobranchaea californica MacFarland (Gastropoda, Opisthobranchia). Z. vergl.
  Physiol. 75: 207-232.

Dawkins, R. and M. Dawkins. 1976. Hierarchical organization
  and postural facilitation: Rules for grooming in flies.
  Anim. Behav. 24: 739-755.

Dingle, H. and R.L. Caldwell. 1969. The aggressive and
  territorial behavior of the mantis shrimp Gonodactylus
  bredini Manning (Crustacea: Stomatopoda). Behaviour 33:
  115-136.

Dudziński, M.L. and J.M. Norris. 1970. Principal components
  analysis as an aid in studying animal behavior. Forma et
  Functio 2: 101-109.

Eaves, K.J. 1972. The multivariate analysis of certain
  genotype-environment interactions. Behav. Gen. 2: 241-244.

Eberhard, W. 1969. Computer simulations of orb-web construc-
  tion. Am. Zool. 9: 220-238.

Ekehammer, B., D. Schälling, and D. Magnusson. 1975. Dimensions of stressful situations: A comparison between a
  response analytical and a stimulus analytical approach.
  Multivar. Behav. Res. 10: 155-164.

Ewing, L.S. 1972. Hierarchy and its relation to territory
  in the cockroach Nauphoeta cinerea. Behaviour 42: 152-174.

Frazier, W.T., E.R. Kandel, I Kupfermann, R. Waziri, and R.E.
  Coggeshall. 1967. Morphological and functional proper-
  ties of identified neurons in the abdominal ganglion of
  Aplysia californica. J. Neurophysiol. 30: 1288-1351.

Gollob, H.F. 1968. Compounding sources of variation in
  factor-analytic techniques. Psychol. Bull. 70: 330-344.

Halliday, T.R. 1976. The libidinous newt: An analysis of
  variation in the sexual behaviour of the male smooth next,
  Triturus vulgaris. Anim. Behav. 24: 398-414.

Hamilton, P.V. and H.W. Ambrose, III. 1975. Swimming and
  orientation in Aplysia brasiliana (Mollusca: Gastropoda).
  Mar. Behav. Physiol. 3: 131-144.

Hazlett, B.A. 1972. Responses to agonistic postures by the
  spider crab Microphrys bicornutus. Mar. Behav. Physiol.
  1: 85-92.

_____. 1974. Field observations on interspecific agonistic
  behavior in hermit crabs. Crustanceana 26: 133-138.

_____ and G.F. Estabrook. 1974a. Examination of agonistic
    behavior by character analysis. I. The spider crab
    Microphrys bicornutus. Behaviour 48: 131-144.
_____ and _____. 1974b. Examination of agonistic behavior by
    character analysis. II. Hermit crabs. Behaviour 49:
    88-110.
Hope, K. 1968. Methods of Multivariate Analysis. Univ. of
    London Press, London.
Huntingford, F.A. 1976a. The relationship between anti-pred-
    ator behaviour and aggression among conspecifics in the
    three-spined stickleback, Gasterosteus aculeatus. Anim.
    Behav. 24: 245-260.
_____. 1976b. An investigation of the territorial behaviour
    of the three-spined stickleback (Gasterosteus aculeatus)
    using principal components analysis. Anim. Behav. 24:
    822-834.
James, F.C. 1971. Ordinations of habitat relationships
    among breeding birds. Wilson Bull. 83: 215-236.
Jennrich, R.I. 1977. Stepwise discriminant analysis. In:
    K. Enslein, A. Ralston, and H.S Wilf (eds.) Statistical
    Methods for Digital Computers. Vol. III. John Wiley &
    Sons, Inc., New York.
Kaiser, H.F. 1958. The Varimax criterion for analytic rota-
    tion in factor analysis. Ed. Psychol. Meas. 19: 413-420.
Kandel, E.F. 1976. Cellular Basis of Behavior. W.H. Free-
    man & Co., San Francisco.
Kupfermann, I. 1974. Feeding behavior in Aplysia: A simple
    system for the study of motivation. Behav. Biol. 10:
    1-26.
_____ and T.J. Carew. 1974. Behavior patterns of Aplysia
    californica in its natural habitat. Behav. Biol. 12:
    317-337.
Landau, H.F. 1951a. On dominance relations and the structure
    of animal societies: I. Effect of inherent character-
    istics. Bull. Math. Biophysics 13: 245-262.
_____. 1951b. On dominance relations and the structure of
    animal societies: II. Some effects of possible social
    factors. Bull. Math. Biophysics 13: 245-262.
_____. 1953. On dominance relations and the structure of
    animal societies: III. The condition for a score struc-
    ture. Bull. Math. Biophysics 15: 143-148.
Leamy, L. 1975. Component analysis of osteometric traits in
    random-bred house mice. Syst. Zool. 24: 176-190.
Maurus, M. and H. Pruscha. 1973. Classification of social
    signals in squirrel monkeys by means of cluster analysis.
    Behaviour 47: 106-190.
Marshall, N.J. 1972. Privacy and environment. Human Ecol.
    1: 93-110.
Montgomery, T.H. 1910. The significance of the courtship and

secondary sexual characteristics of araneads. Am. Nat. 44: 151-177.

Morgan, B.J.T., M.J.A. Simpson, J.P. Hanby, and J. Hall-Craggs. 1975. Visualizing interaction and sequential data in animal behaviour: Theory and application of cluster-analysis methods. Behaviour 56: 1-43.

Morrison, D.F. 1967. Multivariate Statistical Methods. McGraw-Hill, New York.

Nunnally, J. 1962. The analysis of profile data. Psychol. Bull. 59: 311-319.

Overall, J.E. 1968. Cluster oriented factor analyses: Oblique powered vector factor analysis. Multivar. Behav. Res. 3: 479-488.

_____ and S.M. Free. 1972. Multidimensional scaling based on a sub-set of objects or variables. Psychometric Laboratory Reports, Number 30, The University of Texas Medical Branch at Galveston.

_____ and C.J. Klett. 1972. Applied Multivariate Analysis. McGraw-Hill, New York.

_____ and J. Porterfield. 1962. Powered vector method of factor analysis. Psychometrika 28: 415-422.

Rohlf, F.J. 1968. Stereograms in numerical taxonomy. Syst. Zool. 17: 246-255.

_____. 1971. Perspectives on the application of multivariate statistics to taxonomy. Taxon. 20: 85-90.

_____ and J. Kishpaugh. 1972. Numerical Taxonomy System of Multivariate Statistical Programs. The State University of New York at Stony Brook, New York.

_____ and R.R. Sokal. 1958. The description of taxonomic relationships by factor analysis. Syst. Zool. 11: 1-16.

Rovner, J.S. 1966. Courtship in spiders without prior sperm induction. Science 152: 543-544.

_____. 1967. Acoustic communication in the lycosid spider (Lycosa rabida Walckenaer). Anim. Behav. 16: 358-369.

_____. 1971. Mechanisms controlling copulatory behavior in wolf spiders (Araneae: Lycosidae). Psyche 78: 150-165.

_____. 1972. Copulation in the lycosid spider (Lycosa rabida Walckanaer): A quantitative study. Anim. Behav. 20: 133-138.

_____. 1974. Copulation in the lycosid spider Schizocosa saltatrix (Hentz): An analysis of papal insertion patterns. Anim. Behav. 22: 94-99.

Royce, J.R., W. Poley and L.T. Yeudall. 1973. Behavior-genetic analysis of mouse emotionality. I. Factor analysis. J. Comp. Physiol. Psychol. 83: 36-47.

Rummel, R.J. 1970. Applied Factor Analysis. Northwestern Univ. Press, Evanston.

Ruskin, R.S. and C.D. Corman. 1971. A multivariate study of competition in a free operant situation. Psychon. Sci.

23: 251-252.

Schwartz, E.L., A. Ramos and E.R. John. 1976. Cluster analysis of evoked potentials from behaving cats. Behav. Biol. 17: 109-117.

Shepard, R.N., A.K. Romney, and S. Nerlove. 1972. Multidimensional Scaling, Vols. I and II. Seminar Press, London.

Smith, P.K. and K Connolly. 1972. Patterns of play and social interactions in pre-school children. p. 65-96 in N.G. Blurton Jones (ed.) Ethological Studies of Child Behavior. Cambridge Univ. Press, Oxford.

Svendsen, G.E. and K.B. Armitage. 1973. Mirror image stimulation applied to field behavior studies. Ecology 54: 623-627.

Tucker, L.R. 1968. Comments on "Compounding sources of variation in factor-analytic techniques". Psychol. Bull. 70: 345-354.

Wiepkema, P.R. 1961. An ethological analysis of the reproductive behaviour of the Bitterling. Arch. Neerl. Zool. 14: 103-199.

Williams, W.R. and J.M. Lambert. 1959. Multivariate methods in plant ecology. I. Association-analysis in plant communities. J. Ecol. 47: 83-101.

Witt, P.N., C.F. Reed and D.B. Peakall. 1968. A Spider's Web: Problems in Regulatory Biology. Springer-Verlag, Inc., New York.

PREDICTING BEHAVIORAL RELATIONSHIPS

Brian A. Hazlett and Catherine E. Bach

University of Michigan

Abstract: A model of the motivational system of animals is presented. It is assumed that we can represent how motivational levels change and affect one another by a series of simultaneous differential equations and that the organism's total behavioral repertoire can be viewed as an entity. By use of stability analysis we can predict that certain qualitative relationships between motivational classes are possible and other relationships could not lead to the desired kind of system's behavior. In particular, motivational classes must act in multiplicative fashions in order to generate motivational systems which can show variations in motivational levels yet are globally stable.

INTRODUCTION

As animals obtain information about their environment (internal and external) they do not react as simple black boxes, altering output as a direct function of input. Rather this information is processed by the central nervous system in complex ways which are altered in the individual by various kinds of past experience.

One of the complexities which we must face as we try to understand animal behavior is that the different kinds of behavior that ethologists generally recognize are not autonomous, but interact in various ways (Fentress, 1976a; McFarland, 1976). Thus the response a hermit crab shows to an aggressive stimulus is not only a function of the characteristics of that stimulus (Hazlett, 1972), but of the aggressive level of the organism (Hazlett, 1969) <u>and</u> the levels of other kinds of behavioral tendencies. For example, the hunger level of a hermit crab alters its aggressive behavior (Hazlett, 1966), as do the reproductive tendencies of the individual at any particular point in time.

What we wish to outline is an approach by which we may predict general, qualitative relationships between classes of behaviors, as a first step towards a testable general model of behavior. While the actual predictions that can be made at this time are of a very general nature, we hope the methods used may prove to be of some heuristic value as a way of looking at the internal dynamics of behavior---of modeling the infrastructure of behavior.

In this regard, the approach we are utilizing is not a quantitative method (and thus does not fit the title of this volume perfectly), but rather it is a mathematically based qualitative method for trying to make general predictions about behavioral organization. It is assumed that the organization(s) of behaviors in animals are present because they maximize inclusive fitness (Hamilton, 1964) just as the sequences of behavior which result from such an organization contribute to an animal's fitness (Sibly and McFarland, 1976).

I. GENERAL CONSIDERATIONS OF THE METHOD

The method to be used is neighborhood stability analysis (Lewontin, 1969). Since exact solutions of differential equations can usually only be obtained for linear equations, we must look at qualitative behavior in the neighborhood of the equilibrium point(s) in the case of non-linear equations, thus the name 'neighborhood stability analysis'. This method has been widely used in ecology to look at the dynamics of communities as a result of the interactions of the species in the community and to predict how species in a community could interact if the community is to persist over time (i.e., be stable in its species composition) and recover from environmental perturbations. Stability analysis is applicable to any system in which the changes in the components of the system can be described as functions of the values of other components of the system. All the problems inherent in the use of this method in ecology (DeAngelis, 1975; May, 1971, 1974) are problems for its use in behavior (also see McFarland, 1971), but some heuristic value may emerge despite these limitations.

In using the method of neighborhood stability analysis, we are assuming that it is not unreasonable to represent the overall motivational state of an organism at any instant in time by a point in what we shall call behavioral potential space. This space is defined by a series of axes each of which represents the motivational level associated with a class or collection of behavior patterns. The number and delineation of those axes could be determined either by more traditional ethological techniques or by multivariate analyses such as those discussed by Aspey in this volume. The location

of the point in n-dimensional behavioral potential space would
depend upon the relative values of the motivational levels at
that instant in time (Figure 1). Movement of this point
(which will be called the behavioral potential point) would
represent changing values of one or more of the motivational
levels. The behavioral potential space described herein is
analagous to the "command space" of McFarland and Sibly
(1972) and Sibly and McCleery (1976).

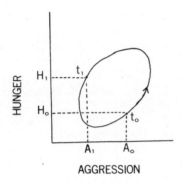

AGGRESSION

*Fig. 1. Phase plane representation of a hypothetical
relationship between aggression and hunger. The indicated
elliptical line represents all possible values of aggression
and hunger as delineated by a set of equations describing the
relationship between hunger and aggression. The levels of
aggression and hunger at any instant in time define the location of the motivation state of the animal in two-dimensional motivational space. At time t_O, the levels of hunger and
aggression are H_O and A_O and similarily for time t_1.*

The purpose of the analysis which follows is twofold.
First, we believe that by the application of systems of simultaneous differential equations it is possible to derive better
descriptive approaches. That is, this approach may provide a
more meaningful method of describing the overall motivation of
organisms. Secondly, the application of stability analysis to
systems of simultaneous differential equations can predict
what kinds of interactions between classes of behavior are
possible. In that regard it is a useful tool in hypothesis
formation. These two functions of this approach are distinct
and this approach is clearly different from any statistical
techniques which deal with questions of probability.
    It is essential that the domain in which we propose to
apply stability analysis is carefully limited. The portion of
the behavioral system modeled in this paper is the overall

motivation of the animal, i.e. the collection of internal rules which modulate input-output. In this initial analysis we are not specifying the nature of either the input functions or the output functions, but looking only at the possible features of the motivational system itself. Clearly the patterns of inputs (stimuli and other sources of environmental modifications) influence an organism's motivational state, just as its own behavior (the output function of the system) will have strong feedback effects on the motivational system. But we wish to suggest that there are features of the motivational system which must exist if it is to work appropriately as it is modified by various inputs and as it in turn modifies behavioral output. It may be that the general statements that are made about the structure of the motivational system in this paper will hold irrespective of the artificiality of trying to look at it in "isolation".

Most of the input functions to animals are not continuous but discrete in time and space --- therefore must be described or modeled by difference equations (as opposed to differential equations). Similarly, the behavioral output functions of animals are discrete. Thus a hermit crab executes aggressive displays at some times and shows feeding behaviors at other times, but does not exhibit a continuum of acts between those classes. This is why discrete functions such as those described by Thom's catastrophe theory (Zeeman, 1976) are logical models for the behavioral acts we observe. However, it seems reasonably clear that the motivational systems of animals can vary continuously (thus be treated by differential equations).

Four basic assumptions in our application of the method of stability analysis are:

(1) The values of the components (motivational axes) being examined can vary over time and how they vary can be represented by a series of simultaneous differential equations. As Maynard Smith (1968) has pointed out, selection of appropriate equations is usually the most difficult step in modeling biological systems. Holling and Buckingham (1976) used an equation similar to that given below in their study of changes in hunger.

(2) The organism should be viewed as having a stable motivational system in which none of the components of the system disappear over time. Further, in any realistic model, the organism should be able to do different things, that is the values of the components can vary. In the terminology of Fentress (1976b), the motivational system which we are considering is the "mechanism" between the causes (environmental inputs) and effects (behavioral outputs) which comprise the total behavioral system.

(3) The behavior of the motivational system (as distinct from the overt behavior of the organism) is controlled primar-

ily by deterministic factors, i.e. that random, stochastic effects are less important. This means that the movement of the behavioral potential state is determined largely by the interactions of the components of the system. (A variate of this approach which incorporates non-deterministic, stochastic effects will be mentioned). Fentress (1976c) argues that this will become more true as behavioral ontogeny occurs.

(4) The results of neighborhood stability analysis are true over the entire range of possible variation in the components, i.e. the system is also globally stable. Neighborhood stability analysis actually looks at the qualitative behavior of the system only in a small area of phase space (in this case behavioral potential space) around the equilibrium values of the system, where the values of the motivational levels are unchanging. It is possible for the systems to behave differently in other regions of the phase space and one model presented later depends upon this characteristic. The existence of global stability can be checked by the calculation of the appropriate Lyapunov functions, but this has not been attempted at this time. If the relationships of the components of a system can be described by linear equations, the results of neighborhood stability analysis will always be true globally.

A. A Simplistic Application

Before proceeding with the mechanics of neighborhood stability analysis, it might be helpful to give some idea of the biological meaning of this procedure by presenting a specific example. This example is clearly overly simplistic, but illustrates what can be accomplished with the method.

Let us assume that we are looking at how two classes of behavioral motivation (hunger and aggression) interact as they change over time. The equations given below say that the change in hunger level over time is a function of the present

$$\frac{dA}{dt} = cA + dH \qquad (1a)$$

$$\frac{dH}{dt} = eA + fH \qquad (1b)$$

hunger level (i.e. the level at time t) and the present aggressive level (each multiplied by a constant which can be either positive, negative, or zero). The constant "f" could represent the rate at which a food-related physiological process proceeds. The change in aggression over time is a function of the present aggressive level and the present hun-

ger level (again multiplied by constants). It should be
emphasized that we are dealing only with variations in behavioral potential level, not the actual probabilities of a behavioral act occurring. Obviously behavioral potential levels
and the probabilities of acts occurring are related, in as yet
unclear ways (see Heiligenberg, 1974, 1976a vs. Houston and
McFarland, 1976), however this need not concern us at this
time.

Given the above equations, we can make some predictions
on how this deterministic system changes. For example, if
the constants c, d, e, and f are positive numbers (of most
values), the system will behave as an unstable node, as shown
in Figure 2. That is, unless the initial levels of aggression

AGGRESSION

Fig. 2. *Pathways of the motivational state in an unstable node system. Any disturbance of the system away from
the indicated equilibrium point is followed by continued
movement of the motivational state in one direction.*

and hunger are at the equilibrium point of the equations, the
system will expand outward (the behavioral potential state
moves away from the starting values and moves continually in
one direction). Even if the system is at the equilibrium
point and gets perturbed in any way, the outward movement proceeds. Eventually the level of one of the classes becomes so
large that the animal is doing only one thing. Clearly this
is not a very useful model if it predicts that hunger level
becomes infinite, either positive or negative. By our earlier
definition, this system is not stable, one of the elements
will functionally disappear.

As an aside, if this 'phase plane' method (as in Figures
1 and 2) of showing how two things behave relative to each
other (with the time axis left out) is unfamiliar, we can
re-illustrate how behavioral levels are changing by plotting
the levels of the behavioral potentials over time. Thus a
system which shows the manner of interaction shown on the left
of Figure 3 is the same as the patterns of change over time
shown on the right of Figure 3.

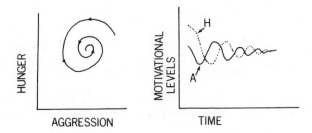

*Fig. 3. The methods of indicating the relationships of
components of a system. The phase plane method on the left
illustrates the same stable oscillatory system as the time
dependent plot on the right.*

Before we can consider the steps involved in neighborhood
stability analysis, a brief background item should be mentioned. Associated with the matrices which describe any set
of equations are numbers called eigenvalues. The meaning of
these numbers may be illustrated by looking at the solutions
for the system of linear differential equations concerning
hunger and aggression presented earlier (equations 1a and 1b).
The solutions for this system can be written in the general
form:

$$A = k_1 e^{\lambda_1 t} + k_2 e^{\lambda_2 t}$$
$$H = k_3 e^{\lambda_1 t} + k_4 e^{\lambda_2 t}$$

where the values λ_1 and λ_2 are the eigenvalues, and k_1, k_2,
k_3, and k_4 are constants. The eigenvalues are usually
written as complex numbers, that is a ± bi. The nature of
these numbers (the real and imaginary parts of the eigenvalues) will tell us about the behavior of the system. In the
above equations, for example, if λ_1 and λ_2 have imaginary
parts, the values of A and H will oscillate over time.
The behavioral state which represents levels of hunger and
aggression will follow an oscillatory path.
    It is interesting to note that such systems would produce
oscillations in motivational levels simply by the ways the
classes of behavior interact. Such oscillations could result
not only from the action of a central rhythmic center
(Pittendrigh and Bruce, 1959) but would be an inevitable result of the way the system is put together. Oscillations may
be a necessary byproduct of having a biologically realistic
system with many parts.

B. Mechanics and Possible Outcomes

To model a system, a series of differential equations are written describing how the components of the system affect one another. Obviously there would be one equation for each component. Neighborhood stability analysis then involves the following steps:

(1) The equilibrium points of the equations are determined and the partial derivatives at the equilibrium points computed.

(2) A matrix of the partial derivatives (called the Jacobian matrix) is constructed.

(3) The eigenvalues for the equations are then obtained by solving the equation Det $|J-\lambda I|=0$, where J is the Jacobian matrix and I is the identity matrix.

(4) A simple table (Table 1) is consulted which shows which of the types of system's behavior applies to the particular system of differential equations, based on the values of the eigenvalues.

TABLE 1

Relationship between eigenvalues and classes of system's behavior. Eigenvalues are written as complex numbers $(a \pm bi)$ where "a" is the real portion and "bi" the imaginary.

Values of a	Value of b	System's Behavior
a_1, $a_2 > 0$	0	Unstable node
$a_1 > 0$, $a_2 < 0$	0	Saddle point
a_1, $a_2 < 0$	0	Stable node
$a_1 = 0$, $a_2 < 0$	0	Neutrally stable point
a_1 and/or $a_2 > 0$	Non-zero	Unstable oscillations
$a_1 = 0$, $a_2 \leq 0$	Non-zero	Neutrally stable oscillations
a_1, $a_2 < 0$	Non-zero	Stable oscillations

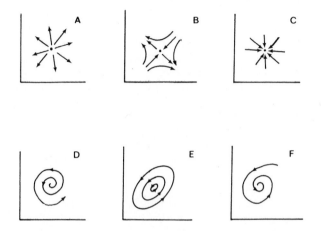

Fig. 4. The six classes of system's behavior discernible by neighborhood stability analysis: (A) unstable node, (B) saddle point, (C) stable node, (D) unstable or undamped oscillations, (E) neutrally stable oscillations or centers and (F) stable or damped oscillations.

A sample of the mechanics of these steps is given for a linear system in Appendix A and for a non-linear system in Appendix B.

There are seven classes of results (and many variations) which can emerge from systems of differential equations, linear and non-linear, when we consider just neighborhood stability. Six of these are shown diagrammatically in Figure 4. Depending on the values of the constants in the linear equations mentioned above, it is possible to obtain all of these classes of system's behavior. Thus depending on how we suggest that aggression and hunger change over time as functions of aggression and hunger, the system will:

(a) always expand outwards in one direction and one of the components of the system is always eliminated. This is the unstable node mentioned earlier. The particular path of outward movement of the behavioral potential state would depend upon the equliibrium point of the equations, the initial values of the motivational axes, and/or the nature of any perturbation of the system. But the system will always behave in this highly unstable manner.

(b) depending on where you start in the system (what the initial values of the motivational levels are) the behavioral potential state will either first converge towards a point (the equilibrium point) and then expand outward or just expand outwards, also eliminating one component of the system. This is termed a saddle point system and is qualitatively as unstable as the previous class.

(c) always, no matter where the system starts, converge on one point (the stable node, equilibrium point of the system), where the levels of the components of the system can no longer change. Thus the motivational levels of the animal could not vary and thus would be functionally non-existent as a means of understanding variations in behavior. This class of system's behavior would be stable but meaningless as a model of animal behavior.

Another class of system's behavior is the special case of a neutrally stable point. This would also be meaningless as a model since no changes in the behavioral potential state would ever occur.

Note that in the above classes of system's behavior, the eigenvalues associated with the equations have no imaginary parts.

(d) always move outward in an expanding spiral, eventually eliminating one of the components. This is called unstable oscillations or unstable focus.

(e) show neutrally stable oscillations (also called centers), in which the components can assume various values, but only certain values. It should be noted that the particular pathway the behavioral potential state follows is set by the initial values of the components --- that is the nature of the equations describing the system determine that it will show neutrally stable oscillations, but whether the oscillations will be large or small depends completely upon where the behavioral potential state was when the system started.

(f) the system will always follow a spiral path inward towards a particular point (the equilibrium point), thus tending to eventually eliminate variability in the values of the components. Such stable oscillations systems present the same problems as a useful model of behavior which class (c) does.

We shall return to one of these classes of system's behavior later, but now propose that the only slightly meaningful pattern of system's behavior for an organism to have is that of centers or neutrally stable oscillations. The components of the system persist over time, yet can have many values and thus the organism can do different things.

However, considering linear models first, neutrally stable oscillations can result from two simultaneous linear equations only if, in equations la and lb given on page 125:

(1) the constant c equals minus f
(2) either d or e is negative
(3) and the absolute value of de is greater than the absolute value of cf. All three of these constraints must be true. Some of the equations which satisfy these constraints are:

$$\frac{dA}{dt} = -2A + 3H \qquad (2a)$$

$$\frac{dH}{dt} = -2A + 2H \qquad (2b)$$

$$\frac{dA}{dt} = 2A - 3H \qquad (3a)$$

$$\frac{dH}{dt} = 2A - 2H \qquad (3b)$$

While the system's behavior resulting from these equations is that of neutrally stable oscillations, the relationships described are unrealistic. The first set (equations 2a and 2b) implies that the present level of aggression always appears to decrease future aggression, while in the second set (equations 3a and 3b) the present hunger level always appears to decrease future hunger. Additional biologically unrealistic models can be generated that will result in neutrally stable oscillations.

Since the "best" deterministic system that can be generated from linear equations suffers from being biologically unrealistic, we can conclude that a linear deterministic system is not useful, remembering that for linear equations neighborhood and global stability are always the same.

C. Alternatives to Linear Deterministic Systems

There are however, two broad categories of systems which look much more reasonable as models of behavioral intrastructure. One is stochastic, non-deterministic modification of deterministic systems. The other category is the formulation of non-linear systems.

Considering the latter category first, one of the many non-linear systems we could formulate, that is one in which the elements interact in their effects, is the following:

$$\frac{dA}{dt} = cA + dAH - eA^2 \qquad (4a)$$

$$\frac{dH}{dt} = fH - gAH \qquad (4b)$$

For some values of the constants in these equations, neutrally stable oscillations result and (unlike the linear models mentioned earlier) the relationships are biologically realistic. That is, the change in aggression is affected not only by the

present state of aggression but also by multiplicative interaction with the present state of hunger. Similarly the
changes in hunger level are a function of the present state of
hunger and an interactive term involving the present state
of hunger and the present state of aggression. The A^2 term
implies that at very high levels of aggression (assuming the
constant "e" is small), a negative feedback effect on aggression occurs. The most interesting outcome of this model is
that the two elements must interact multiplicatively in their
effects on the future levels of the elements. Thus the way
hunger affects aggression depends on the level of aggression.

Even though this model is biologically realistic, two
features of systems showing neutrally stable oscillations need
to be mentioned. First, very small changes in the values of
the constants of the equations will change the system to an
unstable one. That is, such systems are structurally unstable
and there is very little leeway if neutrally stable
oscillations are to result (May, 1974). Secondly, as mentioned earlier, the pathway followed by the behavioral potential state in such a system will remain the same as that set
by the initial conditions until disturbed by outside influences. In this context, the most important outside influences would be stimuli. When a system showing neutrally
stable oscillations is modified by an outside influence, the
behavioral potential state follows a new pathway (again
yielding oscillations of the motivational levels) until disturbed again.

Since we are considering stimuli to act as environmental
perturbations, any acceptable model system has to be able to
accomodate (and work with) these quasi-random inputs. Figure
5 illustrates what would happen when a system of neutrally
stable oscillations is perturbed (affected) by an aggressive
stimulus. The new cycle, with a different amplitude, will now
persist until the animal is again influenced by another outside influence. Clearly outside influences such as stimuli,
climatic changes, and photoperiod changes can result in
changes in behavior and eventually alterations in motivational
levels. But the changes in the amplitude of the oscillations
of the behavioral state illustrated in Figure 5 not only are
unrealistic but could result in negative values for one or
more biological levels.

We could view the effects of past experience as the
shifting of the pathways in a system showing neutrally stable
oscillations, but that would be pushing this approach too
much. Past experience, particularly during the early stages
of behavioral ontogeny, certainly results in modifications of
the behavioral dynamics of an individual. And this should be
reflected in modifications of the equations describing the
interrelationships of the elements of a behavioral system.

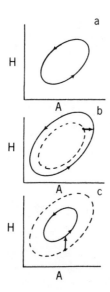

Fig. 5.   Changes in the pathway of a neutrally stable oscillatory system when affected by outside perturbations. The pathway of the motivational state in graph (a) would continue indefinately unless disturbed as in graph (b). Once disturbed, the motivational state would continue in the new pathway (wider oscillations) unless disturbed again (c) when yet another pattern would follow (new smaller oscillations).

However, the kinds of modifications possible in equations which result in neutrally stable oscillations are not very helpful in making such systems biologically realistic.

D.   Global Stability

     Because of the problems mentioned above with systems which yield neutrally stable oscillations (whether by linear or non-linear equations) it should be clear that these systems are not good models. Since the other classes of system's behavior described earlier are even less desirable, it would seem that there is no good deterministic model to be suggested by neighborhood stability analysis. However, there is a class of system's behavior which can result from some sets of non-linear equations which is realistic as a model of behavior but which only emerges when we look at global stability. This is the stable limit cycle. In these systems, as shown in Figure 6, if a perturbation occurs (e.g. a stimulus affects behavior) and the behavioral potential state changes (that is the effects of a stimulus eventually change the value(s) along

Fig. 6.   A system showing a stable limit cycle.   Due to the nature of the relationships of the components, the movement of the motivational state always tends to return to the same pathway regardless of the direction or extent of any perturbation of the system.

one or more motivational axes), the structure of the internal dynamics is such that the system always returns to the same oscillatory path. That is, the pathway which the behavioral potential state follows is set by the nature of the equations, and is not affected by the initial values of the components or by later perturbed values. The system recovers from perturbations as a result of how the classes of behavior affect one another. The permanent oscillations generated by limit cycles are thus much more realistic than those generated by neutrally stable systems. In the case of stable limit cycles, the animal can have a variety of motivational states, show different behavior patterns at the appropriate times, yet maintain stability in its repertoire and the interrelationships of the components of its repertoire in every sense of the word stability.

What sets of equations produce a system which shows stable limit cycles? As developed by Kolmogorov (1936) and explained by May (1974), all formulae of the general form:

$$\frac{dA}{dt} = A \ \ G(H,A) \qquad (5a)$$

$$\frac{dH}{dt} = H \ \ F(H,A) \qquad (5b)$$

where F and G represent functions, will produce either a stable node or a stable limit cycle, providing the equations satisfy the necessary conditions outlined by Kolmogorov (see May, 1974, p. 87). Although the general form of these equations is simple, it is difficult to establish that any specific set of equations satisfies these conditions.

However, we believe it is not necessary to try (at this time) to deal with this mathematical problem in order to make an interesting statement. Based upon the general form of the equations given above we can say that for any deterministic system to show the kind of system's behavior desired, the elements must show multiplicative interaction in their effects. For example, aggression and hunger must interact in some multiplicative manner in their effects on changes in aggression and hunger levels. This is no surprise, is biologically realistic (McFarland, 1976; Houston and McFarland, 1976), and the analysis says that this has to be so.

This requirement of interactive elements in the system is true only for a deterministic system. Jeffries (1974) and Vandermeer (in prep.) have independently shown for ecological systems, that totally random input to what was a stable oscillatory system can result in system's behavior analagous to a stable limit cycle. This is because there will be a greater probability of stochastic factors operating in opposition to deterministic factors due to the shape of the trajectory, especially somewhat near the equilibrium point. Thus even for linear, non-interactive systems, limit cycle behavior can be generated.

In behavior, few inputs will be truly random due to the homeostatic nature of behavior. In any behavioral system which shows unstable oscillations --- that is a system in which the associated eigenvalues have imaginary parts and one or more real parts are positive --- it is possible to have system's behavior that is functionally similar to stable limit cycles if factors not included in the equations are very important. In this context, that means that if the effects of stimuli are quasi-random and strong relative to the effects of the constraints of the behavioral-motivational intra-structure even a linear system could behave in the desired fashion. As the state of the behavioral potential system begins spiralling outward (the pattern of unstable oscillations) the likelihood of an external influence (stimulus) pushing the system inwards increases (see Figure 7). This is because the animal will increase the likelihood of that kind of stimulus reception occurring (either by appetitive movements to a location where the appropriate stimulus object is or by some form of selective attention (Hebb, 1966) or both). Thus the action of the stimuli, influenced by the behavior of the animal, will tend to be in opposition to the path of state movement which would result from the nature of the interrelationships of the system alone, and oscillations similar to a stable limit cycle will result.

CONCLUSION

At this time there is no clear way to decide which of

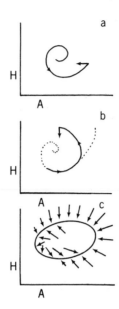

Fig. 7. Generation of system's behavior analagous to a stable limit cycle from strong outside input to an intrinsically unstable oscillatory system. As indicated by the arrows, the motivational state would move in a spiral, expanding path in the regions of lower motivational levels due to the intrinsic relationships and move in a decreasing spiral elsewhere as a result of extrinsic input.

these classes of models might be more biologically realistic. Perhaps some animals have motivational systems which are less easily influenced by outside perturbations and then the system would be more readily described by a series of non-linear equations which produce a stable limit cycle in behavioral potential space just by the way the components of the system interact. Alternatively the pattern of stimulus input and the interaction of the elements could be balanced to result in a functionally useful model of behavioral dynamics.

It should be self-evident that the two component system (hunger and aggression) that we have used for illustrative purposes is not realistic as a real model of behavior. A complete formulation of a realistic model would include a larger number of components (motivational axes) and the influences of outside perturbations on all of the motivational axes of the system. A more complete model would obviously be more powerful in making predictions about the structure of an animal's behavioral repertoire. The mathematics of considering n-dimensional systems (thus building a more complete model of behavior) is difficult but tractable for relatively small num-

bers of dimensions. The results of stability analysis on such
n-dimensional systems would not be easy to determine, but
would represent an intriguing challenge for future work.

As a final comment, it should be pointed out that the
general method of modeling which we have outlined above is an
attempt to look at different aspects of behavior than most of
the other methods discussed in this volume. As such, the
different approaches are complementary to each other rather
than alternative ways of dealing with similar questions. For
example, the clustering techniques examined by Wayne Aspey are
needed to tell us how many axes we should be concerned with in
defining a behavioral potential space, i.e. how many dimensions are there and what are the behavior patterns associated with each axis. Similarly, some form of information
theoretic analysis, as discussed by June Steinberg, can be
very useful as a descriptor of the sequences of behavioral
acts which appear along one or more axes. Or, as has been
done in a few cases, information theory has been used to
measure the relative degree of interaction between classes of
factors affecting behavior (Hazlett and Estabrook, 1974;
Rubenstein and Hazlett, 1974). Information theoretic and
clustering techniques are useful (along with many other statistical approaches) for analysing and describing, post facto,
the behavior patterns we see --- clearly the most important
task at hand for most of us watching animals since variability
is always present in the behavioral data gathered. But
statistical approaches do not lend themselves to the formulation of predictive model building --- a task distinct from
the statistical testing of data or descriptions --- something
we should not ignore.

Behavior always involves the passage of time and
ethologists have often been concerned with the units of time
over which they measure changes. As Schleidt (1973) has
clearly pointed out, behavioral effects do not occur instan-
taneously but over time (also see Heiligenberg, 1976b; Nelson,
1965). While the use of simultaneous differential equations
and stability analysis may not prove to be useful, we should
be cognizant of the general lack of that category of
hypothesis formation in the study of animal behavior. We can
not test the utility or realism of models until they are
devised. We can not take the experimental steps which would
follow from the formulation of a hypothesis or develop better
models until someone is foolish enough to suggest the first
simple model.

Only further analyses and perhaps even some observations
can determine which models are more useful in our attempts to
understand behavior.

ACKNOWLEDGMENTS

We wish to express our thanks to John Vandermeer, Les Real, Ron Westman and John Holland for their helpful criticism. The initial stimulus for the ideas in this paper came from meetings of the mathematical ecology group and were refined by discussions with the theoretical biology group at the University of Michigan.

APPENDIX A

This appendix outlines the steps involved in carrying out neighborhood stability analysis of a generalized system of 2 linear differential equations. The stability of one system of specific equations is then analyzed.

(1) Simultaneous differential equations are written to describe the system.

$$\text{Let} \quad \frac{dA}{dt} = cA + dH \qquad \text{(A-1a)}$$

$$\text{and} \quad \frac{dH}{dt} = eA + fH \qquad \text{(A-1b)}$$

where A = level of aggression, H = level of hunger, and c, d, e, and f are constants.

(2) A matrix of coefficients (M) is constructed. (This is actually a specialized case of the Jacobian matrix of partial derivatives which is constructed for non-linear systems, see Appendix B.)

$$M = \begin{vmatrix} c & d \\ e & f \end{vmatrix}$$

(3) The eigenvalues of the coefficient matrix are now computed by solving the equation:

$$\text{Det}|M - \lambda I| = 0 \qquad \text{(A-2)}$$

where Det = determinant of a matrix, λ = eigenvalue, and I = identity matrix.

$$\text{Det}|M-\lambda I| = \text{Det}\left[\begin{vmatrix} c & d \\ e & f \end{vmatrix} - \lambda \begin{vmatrix} 1 & 0 \\ 0 & 1 \end{vmatrix}\right] = \text{Det}\left[\begin{vmatrix} c & d \\ e & f \end{vmatrix} - \begin{vmatrix} \lambda & 0 \\ 0 & \lambda \end{vmatrix}\right]$$

Thus equation A-2 takes the form:

$$\text{Det} \begin{vmatrix} c-\lambda & d \\ e & f-\lambda \end{vmatrix} = 0$$

Since by definition, the Det $\begin{vmatrix} a & b \\ c & d \end{vmatrix}$ = ad - bc, we now obtain:

$$(c - \lambda)(f - \lambda) - de = 0$$

That is:

$$\lambda^2 - (c + f)\lambda + cf - de = 0$$

Using the quadratic formula, we obtain:

$$\lambda = \frac{1}{2}\left(c+f \pm \sqrt{(c+f)^2 - 4\,(cf-de)}\right) \qquad (A-3)$$

(4) The configuration of the eigenvalues now tell us about the stability of this system of linear differential equations (see Table 1).

Example

The following system corresponds to equations A-la and A-1b, with c = -2, d = 5, e = -1, and f = 2.

$$\frac{dA}{dt} = -2A + 5H$$

$$\frac{dH}{dt} = -A + 2H$$

Substituting these values of c, d, e, and f into equation A-3, we obtain:

$$\lambda = \frac{0 \pm \sqrt{0^2-4(-4-(-5))}}{2} = 0 \pm \sqrt{\frac{-4}{2}} = \pm i$$

By consulting Table 1, it can be seen that this system exhibits neutrally stable oscillations (case e, Figure 4).

APPENDIX B

This appendix outlines the steps involved in carrying out neighborhood stability analysis of a system of 2 non-linear differential equations. The stability of one system of specific equations is then analyzed.

(1) Differential equations are written to describe the system.

$$\text{Let } \frac{dA}{dt} = f_1(A,H) = cA + dAH - eA^2 \qquad \text{(B-1a)}$$

$$\text{and } \frac{dH}{dt} = f_2(A,H) = fH - gAH \qquad \text{(B-1b)}$$

where A = level of aggression, H = level of hunger, and c, d, e, f, and g are positive constants.

(2) The equilibrium points are determined by setting equations B-1a and B-1b equal to zero and solving the resulting equations (i.e. the system is at equilibrium when there is no change in aggression or hunger levels over time). Note that neighborhood stability analysis assumes linearity in the vicinity of the equilibrium point.
So we obtain:

$$0 = \frac{dA}{dt} = cA + dAH - eA^2 \qquad \text{(B-2a)}$$

$$\text{and } \quad 0 = \frac{dH}{dt} = fH - gAH \qquad \text{(B-2b)}$$

Rearranging terms gives:

$$0 = A (c + dH - eA)$$
$$0 = H (f - gA)$$

The equilibrium levels of aggression and hunger (represented by A* and H*) are non-zero solutions to equations B-2a and B-2b, which clearly gives:

$$c + dH - eA = 0 \qquad \text{(B-3a)}$$
$$\text{and } \quad f - gA = 0 \qquad \text{(B-3b)}$$

Solving equation B-3b, one obtains A* = f/g

Now substituting the value of A* into equation B-3a gives:

$$c + dH - e(f/g) = 0$$

Rearranging and simplification gives:

$$dH = ef/g - c = \frac{ef - cg}{g}$$

Thus:

$$H* = \frac{ef - cg}{dg}$$

(3) The partial derivatives of equations B-1a and B-1b, F_{11}, F_{12}, F_{21}, and F_{22} are evaluated at the equilibrium point. (For non-linear systems these partial derivatives are not constants as in linear systems, and thus must be evaluated at equilibrium.)

$$F_{11} = \left(\frac{\partial f_1}{\partial A}\right)^* \quad F_{12} = \left(\frac{\partial f_1}{\partial H}\right)^* \quad F_{21} = \left(\frac{\partial f_2}{\partial A}\right)^* \quad F_{22} = \left(\frac{\partial f_2}{\partial H}\right)^*$$

Calculating the partial derivative for F_{11} gives:

$$F_{11} = (c + dH - 2eA)^*, \text{ where * means evaluated at equilibrium values of A and H.}$$

Substituting in the values of A* and H*, we obtain:

$$F_{11} = c + d\left(\frac{ef-cg}{dg}\right) - 2e(f/g)$$

Further algebraic simplification gives:

$$F_{11} = -ef/g$$

In a similar manner, we obtain:

$$F_{12} = (dA)^* = d(f/g) = df/g$$
$$F_{21} = (-gH)^* = -g\left(\frac{ef-cg}{dg}\right) = \frac{cg-ef}{d}$$

$$F_{22} = (f - gA)^* = f - g(f/g) = 0$$

(4) A Jacobian matrix of partial derivatives is now constructed, where:

$$J = \begin{vmatrix} F_{11} & F_{12} \\ F_{21} & F_{22} \end{vmatrix} = \begin{vmatrix} \dfrac{-ef}{g} & \dfrac{df}{g} \\ \dfrac{cg-ef}{d} & 0 \end{vmatrix}$$

(5)   The eigenvalues for the matrix J are calculated by solving the equation Det $|J-\lambda I| = 0$, where λ = eigenvalue and I = identity matrix.

Thus:

$$\text{Det}|J-\lambda I| = \text{Det} \left(\begin{vmatrix} F_{11} & F_{12} \\ F_{21} & F_{22} \end{vmatrix} - \lambda \begin{vmatrix} 1 & 0 \\ 0 & 1 \end{vmatrix} \right) = \text{Det} \begin{vmatrix} F_{11}-\lambda & F_{12} \\ F_{21} & F_{22}-\lambda \end{vmatrix} = 0 \quad (B-4)$$

Substituting in the values of F_{11}, F_{12}, F_{21}, and F_{22} into equation B-4 gives:

$$\text{Det} \begin{vmatrix} \dfrac{-ef}{g}-\lambda & \dfrac{df}{g} \\ \dfrac{cg-ef}{g} & -\lambda \end{vmatrix} = 0$$

That is:

$$\left(-\frac{ef}{g}-\lambda\right)(-\lambda) - \left(\frac{cg-ef}{d}\right)\left(\frac{df}{g}\right) = 0$$

Further simplification leads to:

$$\lambda^2 + \frac{ef}{g}\lambda - \frac{f}{g}(cg - ef) = 0$$

Using the quadratic formula, we obtain:

$$\lambda = \frac{1}{2}\left(-\frac{ef}{g} \pm \sqrt{\left(\frac{ef}{g}\right)^2 + \frac{4f}{g}(cg - ef)}\right) \quad (B-5)$$

The following formula enables the calculation of the eigenvalues without going through all of the above steps:

$$\lambda = \frac{1}{2}\left(F_{11} + F_{12} \pm \sqrt{(F_{11} + F_{22})^2 - 4 \det J}\right)$$

(6)   Check the stability of the system in the same way as described in step 4 of Appendix A.

Example

The following system corresponds to equations B-1a and B-1b, with c = 1, d = 1, e = -.5, f = 4, and g = -1.

$$\frac{dA}{dt} = A + AH - .5A^2$$

$$\frac{dH}{dt} = 4H - AH$$

Substituting these values of c, d, e, f, and g into equation B-5 gives:

$$\lambda = \frac{1}{2} (-2 \pm \sqrt{-12} \,) = -1 \pm \sqrt{3}\; i$$

By consulting Table 1, it can be shown that this system of equations behaves in a stable oscillatory fashion ($a<0$ and $b \neq 0$).

REFERENCES

DeAngelis, D.L. 1975. Estimates of predator-prey limit cycles. Bull. Math. Biol. 37: 291-299.

Fentress, J.C. 1976a. Dynamic boundaries of patterned behaviour: interaction and self organization. p. 135-169 in P.P.G. Bateson and R.A. Hinde (eds.) Growing Points in Ethology, Cambridge Univ. Press, London.

_____. 1976b. Behavioral networks and the simpler systems approach. p. 5-20 in J.C. Fentress (ed.) Simpler Networks and Behavior, Sinauer Assoc., Sunderland, Mass.

_____. 1976c. System and mechanism in behavioral biology. p. 330-340 in ibid.

Hamilton, W.D. 1972. Altruism and related phenomena, mainly in social insects. Ann. Rev. Ecol. System. 3: 193-232.

Hazlett, B.A. 1966. Factors affecting the aggressive behavior of the hermit crab Calcinus tibicen. Zeit. f. Tierpsych. 23: 655-671.

_____. 1969. Individual recognition and agonistic behavior in Pagurus bernhardus. Nature 222: 268-269.

_____. 1972. Stimulus characteristics of an agonistic display of the hermit crab (Calcinus tibicen). Anim. Behav. 20: 101-107.

_____ and G. Estabrook. 1974. Examination of agonistic behavior by character analysis. I. The spider crab Microphrys bicornutus. Behaviour 48: 131-144.

Hebb, D.O. 1966. A Textbook of Psychology. Saunders, Phil.

Heiligenberg, W. 1974. Processes governing behavioural states of readiness. in D.S. Lehrman, R.A. Hinde, and E. Shaw (eds.) Advances in the Study of Behaviour, Vol. V, Academic Press, N.Y.

_____. 1976a. The interaction of stimulus patterns controlling aggressiveness in the cichlid fish Haplochromis burtoni. Anim. Behav. 24: 452-458.

_____. 1976b. A probabilistic approach to the motivation of behavior. p. 301-313 in J.C. Fentress (ed.) Simpler

Networks and Behavior, Sinauer Assoc., Sunderland, Mass.
Holling, C.S. and S. Buckingham. 1976. A behavioral model
    of predator-prey functional responses. Behavioral Sci.
    21: 183-195.
Houston, A. and D. McFarland. 1976. On the measurement of
    motivational variables. Anim. Behav. 24: 459-475.
Jeffries, C. 1974. Probabilistic limit cycles. p. 123-131 in
    P.V.D. Driesche (ed.) Proceeding of Victoria Conference on
    Mathematical Problems in Biology, Springer-Verlag, Berlin.
Lewontin, R.C. 1969. The meaning of stability. Brookhaven
    Symposia Biol. 22: 13-24.
Kolomogorov, A.N. 1936. Sulla Teoria di Volterra della Lotta
    per l'Esisttenza. Giorn. Instituto Ital. Attuari 7:
    74-80.
May, R.M. 1971. Stability in model ecosystems. Proc. Ecol.
    Soc. Australia 6: 18-56.
_____. 1974. Stability and Complexity in Model Ecosystems.
    Princeton Univ. Press, Princeton, 265 p.
Maynard Smith, J. 1968. Mathematical Ideas in Biology.
    Cambridge Univ. Press, Cambridge, 152 p.
McFarland, D.J. 1971. Feedback Mechanisms in Animal
    Behaviour. Academic Press, N.Y.
_____. 1976. Form and function in the temporal organization
    of behaviour. p. 55-93 in P.P.G. Bateson and R.A. Hinde
    (eds.) Growing Points in Ethology, Cambridge Univ. Press,
    Cambridge.
_____ and R.M. Sibly. 1972. 'Unitary drives' revisited.
    Anim. Behav. 20: 548-563.
Nelson, K. 1965. After-effects of courtship in the male
    three-spined stickleback. Zeit. vergl. Physiol. 50:
    569-597.
Pittendrigh, C.S. and V.C. Bruce. 1959. Daily rhythms as
    coupled oscillator systems and their relation to
    thermoperiodism and photoperiodism. p. 475-505 in
    Photoperiodism and Related Phenomena in Plants and Anim-
    als, AAAS, Washington.
Rubenstein, D.I. and B.A. Hazlett. 1974. Examination of the
    agonistic behavior of the crayfish Orconectes virilis by
    character analysis. Behaviour 50: 193-216.
Schleidt, W.M. 1973. Tonic communication: Continual effects
    of discrete signs in animal communication systems. J.
    Theor. Biol. 42: 359-386.
Sibly, R.M. and R.H. McCleery. 1976. The dominance boundary
    method of determining motivational state. Anim. Behav.
    24: 108-124.
_____ and D. McFarland. 1976. On the fitness of behavior
    sequences. Am. Nat. 110: 601-617.
Zeeman, E.C. 1976. Catastrophe theory. Sci. Amer. 234 (4):
    65-83.

ENVIRONMENTAL LANGUAGES AND THE
FUNCTIONAL BASES OF ANIMAL BEHAVIOR

Ronald S. Westman

The University of Michigan

Abstract: Traditional methods of analyzing behavioral sequences are reviewed and found to be less adequate to deal with a number of important aspects of animal behavior than are linguistic models.   Linguistic models form both a biologically interesting and mathematically tractable basis for describing animal-environment systems.   Therefore, following an overview of Chomsky's (1956, 1959) phrase-structure grammars, automata are introduced and their relation to languages discussed; this leads to a basic linguistic model of animals.

Various generalizations of languages and grammars are considered in order to produce the basic model which meets the objections raised earlier in discussing other approaches.   It is then shown how various biological considerations such as achieving optimal complexity and minimizing the evolutionary costs of errors serve to delimit the class of acceptable models.   Means of computing the relevant costs and benefits of different strategies are discussed and the problem of fitting linguistic models to data is considered.   Finally, it is suggested that the methods proposed in this paper lead to direct explanations of various interesting biological phenomena.

"It is easy to imagine a language consisting only of orders and reports in battle.——Or a language consisting only of questions and expressions for answering yes or no. And innumerable others.——And to imagine a language means to imagine a form of life."

(Wittgenstein, 1953: par.19)

INTRODUCTION

A.   What is a Language?

Investigators of animal communication often take great
pains to argue that human language is unique, that no other
species possesses a communication mechanism with all the attributes of human language. Hockett and Altmann (1968), for
example, have developed a list of 16 "design features" characterizing human language; they point out that at least one of
these factors is lacking in each non-human communication system (see also Thorpe, 1972 for further discussion in relation
to experiments concerned with teaching human language to chim-
panzees). The argument has been even more strongly stated by
Lenneberg (1964): "There is no evidence that any nonhuman form
has the capacity to acquire even the most primitive stages of
[human] language development."
There may be many reasons why people feel uncomfortable
with the notion of animal languages. For one, we are heir to
the arguments as crystalized by Descartes that only humans
have souls and that other animals, being mere automata, are
at a fundamentally lower order (Boring, 1950). Second, the
idea that animals possess language strikes many people as
anthropomorphic, given the pervasiveness of human-like and
articulate animals in our literature and culture.
More important, however, is the fact that the notion of
language usually is not totally appropriate. Wittgenstein
(1953) has argued in essence that language-games, i.e., the
activities of defining, learning and using language, are ways
of life and are governed by their own sets of rules. Furthermore, the specific words used are defined by the ways in which
they are used rather than by fixed relations to specific objects which they represent.
These ideas can be extended to encompass behavior if we
consider the acts of animals the operational equivalent of
words. More specifically, if we suppose that behavioral acts
admit to classification and groupings when considered from the
animal's point of view, then activities constitute linguistic
behavior in Wittgenstein's sense. Furthermore, the ultimate
meaning of the animal's behavior can be found only by considering its ecological and evolutionary effect in conjunction
with more immediate consequences. Since the animal's behavior
is conditioned by and responsive to events occurring in its
environment, behavior is simply the animal's side of an on-
going operational dialogue between the animal and its environment. In this sense social communication is only one of the
language-games in which an animal engages.
It is to be understood that the term "behavior" in this

context includes not only overt ethological acts but physiological and developmental events as well. All too often practical restrictions on empirical methods have led to the study of
development, physiology and ethology separately and without
adequate reference to each other. Nevertheless it is necessary to retain the distinctions in part if we are to have any
hope of deriving an empirical methodology from the theory presented here. In doing so, however, it must be remembered that
the distinctions are provisional and for methodological convenience only.

B. Scope of This Paper

     I shall argue below that a linguistic approach not only
leads to viable experimental methods but permits inclusion of
many considerations not encompassed by traditional methods,
such as effects of the spatial structure of an animal's environment or the import of variable durations of and time intervals between events. Because good science rests on a
foundation of good empiricism I have chosen to restrict this
paper to operational methods for modelling animal-environment
systems. However, this emphasis should not obscure the fact
that the linguistic point of view presented here leads to
ways of conceptualizing and interpreting animal behaviors that
differ substantially from those currently in fashion.
     I will first briefly review (Section I) the currently
utilized methods for describing animal behavior. In the process a number of "desiderata" are derived on the basis of deficiencies of the current methods and other considerations. In
Section II, I discuss the basic mathematical concepts and constructs to be used in deriving a basic formalism for describing animal behavior. In Section III, I show how the basic
framework can be extended to incorporate the desiderata derived in Section I. Section IV is devoted to showing how evolutionary considerations provide us with the the necessary criteria for selecting amongst the various models which our general method can provide. Finally, I present a brief overview
of evidence supporting the conceptual framework presented here.
     Those familiar with language theory will note that I have
restricted my treatment to phrase-structure grammars and languages. The reader should be aware that alternative approaches to formulating languages exist, for example, the so-called
transformational grammars (Burt, 1971; Chomsky, 1957, 1965).
These other approaches have not been particularly tractable to
mathematical analysis, and, hence, have led to relatively little application outside the field of human linguistics.
     It is crucial to distinguish between the models presented
here and the theory underlying those models. I do not argue

that animals are, or even behave like, automata. I propose
these models because they intuitively seem to represent the
theory much better than any other class of models, because
they possess a number of highly appropriate mathematical properties, and because they can be applied fairly directly to
empirical work.

I. DESCRIPTIVE PARADIGMS IN ANIMAL BEHAVIOR

The first phase of behavioral analysis consists of the
breakdown of an observed stream of behavior into component
units which are individually recognizable and of erecting a
classification scheme to group the behavioral units by common
properties (Hinde, 1970). In general, behavioral biologists
use one or both of two descriptive methods: 1) description
in terms of the patterns and degrees of muscular movements,
glandular activities, etc. comprising the behavior, and 2)
description in terms of consequences, i.e., the results of the
activity.

Several methods of classification have been proposed.
Hinde (1970) recognizes three major schemes. 1) Classification in terms of immediate causation. Such causation may be
due to external stimuli, e.g., social behaviors in response
to the presence of others of one's own species, or internal
stimuli e.g., elevated level of sex hormones. 2) Classification in terms of function, e.g., behaviors leading to the in-
gestion of food would be termed "feeding behaviors."
3) Classification by origin. This may refer either to the
evolutionary origin (and hence, indirectly, to adaptive significance) or to mode of acquisition, e.g., whether learned,
genetically encoded, etc. While these three schemes ignore
the behavioral contexts in which the behaviors appear, other
methods have been employed which do use contextual information
in classification. Cluster-analytic methods can be used to
bring together behaviors occurring in temporal proximity
(Maurus and Pruscha, 1973; Morgan and Simpson, 1976). Furthermore, behaviors tending to occur in the same contexts may
also be considered as similar (Dawkins, 1976; Maurus and
Pruscha, 1973). Clearly other classification schemes can be
used.

A. Methods of Analyzing Behavioral Sequences

Turning from single events to entire sequences we find
again a wide variety of techniques in use. Several investigators have used Markov chains to model behavior sequences
(Bolles, 1960; Chatfield and Lemon, 1970; Fentress, 1972;

Johnson and Hubbell, 1974; Lemon and Chatfield, 1971; Nelson, 1964). In two cases (Lemon and Chatfield, 1971; Nelson, 1964) a first-order model was found to be satisfactory, but this is by no means generally true (Slater, 1973). Unfortunately, Markov chain techniques require that the behavioral process be stationary and typically it is not. Furthermore, it is very difficult to distinguish between rare and forbidden transitions empirically without an enormous amount of data. Finally, when durations are variable, it may be difficult to determine when a behavior is repeated as opposed to being continued. Some investigators have developed related methods for analyzing behavior sequences, such as character state analysis (Hazlett and Estabrook, 1974a,b) and "jackknifing" (see Steinberg, this volume). Since these methods have so far found their usefulness in analyzing social interactions, they fall somewhat outside the scope of this paper but they show some promise for dealing with other, non-social, forms of behavior.

Another technique for analyzing behavioral sequences consists of examining the autocorrelation within and cross-correlations between behavioral sequences (Campbell and Shipp, 1974; Delius, 1969; Heiligenberg, 1973, 1976). In this form, the rate of occurrence of each behavior is considered as a (continuous) stochastic process, and frequency-domain methods are used to study the temporal structure of each process. Two general types of information emerge from such analyses. 1) Frequency-domain analyses will show periodicities and can be used to estimate the degree of clumping of occurrences of the behavior (Campbell and Shipp, 1974). 2) Time-domain analyses will show positional effects of one behavior on another, i.e., to what degree probability of occurrence of a given behavior is elevated or depressed as one moves along the time axis on either side of another behavior (Heiligenberg, 1976). Such methods share with Markovian analyses the requirements that the processes be stationary. (While analytic methods for non-stationarity time series are available (Box and Jenkins, 1976; Kashyap and Rao, 1976) they have not yet found application in studies of animal behavior.) They are also unsuitable for studying relationships between three or more behaviors.

In contrast to Markov and time-series methods which bring out sequence and periodic effects, multivariate methods, such as factor analysis, have been used to try to determine motivational states (Aspey, this volume; Baerend et al., 1970; Van Hooff, 1970; Wiepkema, 1961). The essence of factor analysis lies in the extraction of a relatively small number of variables, or factors, which if acting as causes, would explain most of the observed correlations between behaviors. However, this method also suffers from disadvantages. Se-

quence effects are assumed to be unimportant (Slater and
Ollason, 1972). Furthermore, the "factors" are often diffi-
cult to interpret biologically. For a fuller discussion, see
Slater, 1973.

Syntactic methods, i.e., methods which analyze sequences
as if they were sentences in a language and could therefore
be analyzed grammatically, have also been proposed and used.
The first user seems to have been Marshall in an unpublished
paper in 1965 (see discussions of this paper in Dawkins, 1976;
Hutt and Hutt, 1970; and Vowles, 1970). His grammar was proposed to generate the sequences of pigeon courtship described
by Fabricius and Jansson (1963). Although Marshall was able
to show that his grammar could account for many of the findings of Fabricius and Jansson, it should be noted that he was
describing only one category of behavior in the animal.

Although the analogy between behavior sequences and
grammatical models had been independently generated by Kalmus
(1969), it was not until Fentress and Stilwell studied face-
grooming behavior in mice that grammatical structure in a
data set was pointed out again (Fentress, 1972, 1967a; Fentress and Stilwell, 1973). Fentress has argued in these papers that, while face-grooming sequences in adult mice can be
reliably divided into seven or eight components following one
another by statistically determined rules, these movement components can be best described in relation to each other if
they are grouped hierarchically rather than being subjected to
standard stochastic analysis, because given elements partici-
pate in the formation of different clusters. However, Fentress has not yet formalized these notions into an actual
grammatical representation of his data.

B. Limitations of These Methods

The methods of analyzing behavioral sequences described
above share several serious limitations. First, most methods
focus on the observed behavior of the animal without giving
any consideration to external causative factors; even the factor analytic techniques do not explicitly recognize the
difference between endogenous and exogenous factors. While
the correlative techniques do permit some description of relations between environments and behavior for the animal, they
do not elucidate causal mechanisms. Furthermore, since the
internal state of the organism is not considered, there is no
way to evaluate the consequences for the animal of its own
behavior. As a result, no evolutionary analysis can be performed and the models are therefore somewhat deficient
biologically.

Second, the methods fail to account for the spatial be-

havior of animals. Numerous articles have argued for a connection between the abundance and distribution of resources in
space on the one hand and the movement pattern of a foraging
animal on the other (e.g., Estabrook and Dunham, 1976; Kiester
and Slatkin, 1974; Pulliam, 1974; Schoener, 1971). Numerous
investigators have found that animals adjust their foraging
behavior to variations of resource abundance in space and time
(Alcock, 1973; Goss-Custard, 1970; Krebs, 1973; Murdoch and
Oaten, 1975; Royama, 1970; Slater, 1974; Smith, 1974a,b;
Smith and Sweatman, 1974). It has so far, however, proven
difficult to quantitatively evaluate just how successful the
animals are at optimizing their own behavior. One reason for
this lies in the fact that few of the descriptive schemes for
spatial distributions (see, for example, Dacey, 1963a,b, 1965,
1969; Huijbregts, 1975; Pielou, 1969, 1975; Robinson, 1975;
Tobler, 1975) have been applied or, for that matter, are
likely to be applicable (see below for further discussion of
this point).

Third, the methods outlined do not deal adequately with
the temporal organization of behavior. Correlational methods
tend to either break time up into units of equal length or to
treat it as a continuous variable, thus ignoring the discrete-
ness of behavioral events. Pure Markovian models assume a
uniform time base. While semi-Markov models have been suggested (Altmann, 1974; Cane, 1959), they have so far found little
application. Yet Fentress (1976) states that the duration of
licking movements (in mouse face-grooming) is an important
indicator as to whether movements defined as single strokes or
overhands are most likely to follow. Furthermore, Schleidt
has argued that essential aspects of communication processes
are lost if only the sequence is considered and actual durations are omitted (Schleidt, 1973), and Fentress has made
similar arguments vis-a-vis behavior sequences in general.

Fourth, the methods discussed above fail to account for
the behavior of the organism in physical terms. While a great
deal of attention has been devoted to conceptual models coupling stimulus sets to response sets (for reviews, see Garcia
et al., 1973; Hinde, 1970), no widely applicable quantitative
paradigms have yet emerged. On the other hand, there is a
wealth of data and theory at the physiological and neurophysiological level concerning the logical structure and function of behavioral systems (Conrad et al., 1974; Fentress,
1976; Hinde, 1970; Uttal, 1972). We are at the stage where,
using the terminology of Klir (1969), we have been concen-
trating on examining the permanent behavior of animals without
giving adequate consideration to either the state-transition
structure or to the components and interconnections constituting the animal, much less to the inter-relations between
these three definitions.

Finally, only the grammatical approach allows explicit recognition of hierarchical organization within behavior. In a recent review article Dawkins has amassed an impressive amount of theoretical and empirical evidence for the existence of hierarchical organization in behavior sequences (Dawkins, 1976). In a wider sense, it can be argued that the various response mechanisms of an animal form a hierarchy, with the most rapid (and energetically most expensive) responses being ethological, and, in order of lessening speed and cost, physiological, acclimational, developmental/morphological and finally, in a population sense, genetical responses (see also Slobodkin, 1968). Furthermore, both neural and endocrine systems exhibit hierarchical structure (Gordon et al., 1972; Hinde, 1970). There are also a number of studies indicating the importance of hierarchical structure in the environment (Gibson, 1966; Reed, 1973; Root, 1975).

II. SYNTACTICAL STRUCTURES

The term "syntax" refers to the rules, or principles, whereby a set of words is arranged to convey a particular meaning. As such, the role of syntax is to bring out the intended semantic content. Writing "man dog the mad the bites" conveys little meaning, whereas "the mad man bites the dog" conveys considerable meaning. The words in the two strings are the same, but the relative placements are not.

Sets of events or behaviors can also show distinctive syntactical organization. The sequence "close door; walk through doorway, open door" is nonsensical, not because the individual acts are impossible, but because the set of events indicated cannot happen in that order. Similarly changing the order of the acts in a display sequence or the order in which the elements of a bird's song appear may radically alter or destroy their meanings (Nelson, 1973).

The syntactic aspects of a language are typically encoded in a grammar. While the notions of language and grammar have been around for a long time, useful mathematical models of them have become available only recently. The fundamental formulations are due to Chomsky who originally proposed and classified a number of grammatical models (Chomsky, 1956, 1959). Subsequently these models have been developed extensively and are widely used in a number of areas (Arbib, 1969; Fu, 1974; Hopcroft and Ullman, 1969). The general class to be considered here is the so-called phrase-structure grammar and its generalizations.

In the present section I will discuss the relatively simple basic models of phrase-structure grammars and the equivalent automata. The grammar may be conceptualized as a device

for <u>generating</u> sentences while automata are devices for
<u>decomposing</u> sentences to see if the sentences fit into their
language. In a few words, grammars talk to automata, and
automata understand grammars if and only if they share a common language. In Section III I shall develop both automata
and grammars whose languages more closely approximate those
characterizing real animals in real environments.

A. Phrase-Structure Grammars

    A phrase-structure grammar (or grammar, for short) is a
collection of symbols and rules for how to put the symbols together to make strings of symbols. The rules are formally
known as <u>productions</u>, the terms "rule" and "production" will
be used interchangeably. Two types of symbols are recognized.
Symbols which actually occur in the finished sentences form
the "vocabulary" of the language and are referred to as
<u>terminal symbols</u>. A second class of symbols is included, each
of which represents a whole group of terminal symbols; these
permit delaying the ultimate choice of terminal symbols as
long as possible and thereby markedly reduce the number of
rules required in the grammar. Symbols of this second type
are known as <u>nonterminal symbols</u>. One of the nonterminal
symbols is designated as the start symbol; from it are derived
all sentences in the language through successive substitutions
according to the productions in the grammar.
    More formally, a grammar is a set G containing three sets
and a single (nonterminal) symbol, i.e.,

$$G = \langle V_T, V_N, P, S \rangle$$

where V_T = the set of terminal symbols, V_N = the set of non-
terminal symbols, and P = the set of productions. In general,
these productions serve to (a) generate an initial string of
nonterminal symbols and (b) replace nonterminal symbols by
other strings of symbols, including terminal symbols, and S =
the start symbol. The set of all strings comprised <u>only</u> of
terminal symbols generated by the productions of G from the
start symbol S is called the <u>language</u> of G.

B. Examples

    We start with a simple grammar modelling strings of
behaviors formed of only two types of behaviors. They will be
labelled "0" and "1".

    Ex. 1. Let G_1 be the following grammar:

$$G_1 = \langle V_T, V_N, P, S \rangle$$

where V_T = {0,1} (this notation means that the set V_T contains only 0 and 1);

V_N = {S,A,B}

and P contains the following 7 productions:

1) S $\longrightarrow$ A (this means S can be replaced by A wherever S occurs);

2) A $\longrightarrow$ 1B (this means A can be replaced by the pair of symbols 1B wherever A occurs);

3) A $\longrightarrow$ 0B

4) 0B $\longrightarrow$ 01A (this means B can be replaced by 1A only when the B is immediately preceded by 0. Hence this is a context-sensitive rule.

5) 1B $\longrightarrow$ 10A

6) 1B $\longrightarrow$ 10

7) 0B $\longrightarrow$ 01

There is a large number of "sentences" which can be generated using this grammar. For example:

Step 1:
    S $\longrightarrow$ A (Rule 1). Since we must start with S and rule 1 is the only production with S on the left-hand side, we are forced to use it.

Step 2:
    A $\longrightarrow$ 1B (Rule 2). The choice of production here is arbitrary. We could have used Rule 3 equally well.

Step 3:
    1B $\longrightarrow$ 10A (Rule 5). We could have used Rule 6 instead; that would have ended the derivation with the sentence "10".

Step 4:
    10A $\longrightarrow$ 100B (Rule 3). Again, we could have used rule 2.

<u>Step</u> <u>5</u>:

     100B———>1001 (Rule 7). We could have built a
longer string if we had used Rule 4. The string "1001"
is a sentence in our language, since it contains no non-
terminal symbols.

The language generated by G_1 consists of all strings
formed of mixes of the substrings 01 and 10, such as "10",
"01", "0101", "101001101010010110", etc.

Note that the terminal symbols "0" and "1" could themselves stand for other things; they might, for example, stand
for "foraging behavior" and "aggressive behavior". In
essence the grammar is a <u>syntactical</u> form that exists without
reference to <u>semantical</u> considerations. On the other hand, a
fair amount of semantic structure can be reflected in the
syntax of a language. This point will be discussed below.

Note also that the structure of these sequences cannot
be represented by a simple Markov model. If the elements are
numbered from left to right, the even-numbered elements are
completely predictable on the basis of the odd-numbered elements to their immediate left, but the odd-numbered elements
cannot be predicted on the basis of their even-numbered
predecessors. On the other hand, lumping each even-numbered
element with its immediate predecessor to form strings whose
terminals are "01" and "10" leaves one with an independent-
trials process. Furthermore, the process generating these
strings is non-stationary, since the transitions from odd-
to even-numbered elements obey different rules than the transitions from even- to odd-numbered steps.

Let us now consider a couple of examples exhibiting more
complex behavior. The languages generated by both of the
following grammars are cyclic but in different ways. First,
we shall model a set of sequences with a fixed cycle length
but variability in what state appears in each portion of the
cycle; this type of model lends itself to describing diurnal
or seasonal variations. Second, we shall model the
situation in which the sequence of states is fixed but the
length of time spent in each state is variable; such a model
could be used to model courtship or recognition display sequences.

    Ex. 2.

      (a) For the fixed-cycle-length case,

$$G_{2A} = \langle V_T, V_N, P, S \rangle$$

where $V_T = \{0,1,2,3\}, \quad V_N = \{S,A,B,C\}$

and P contains the following productions:

1) $S \longrightarrow A$	5) $A \longrightarrow 0$	8) $B \longrightarrow 2$	
2) $A \longrightarrow AB$	6) $A \longrightarrow 1$	9) $C \longrightarrow 2$	
3) $B \longrightarrow BC$	7) $B \longrightarrow 1$	10) $C \longrightarrow 3$	
4) $C \longrightarrow CA$			

The sentences produced by G_{2A} will be cyclic with a cycle length of three states; the first state of a triplet will be either 0 or 1, the second 1 or 2 and the third 2 or 3, each with equal likelihoods. The sentences of the language generated by G_{2A} will be strings of such triplets with the last triplet possibly being incomplete; for example

1)   012012012012
2)   113023012122022
3)   1230121131

As indicated previously this type of grammar can be used to model seasonal processes. The fineness of resolution in time may be increased by increasing the number of states per cycle; the fineness of resolution in state may be improved by expanding V_T and, as a result, P. Serial correlation could be accounted for by incorporating context-sensitive rules, e.g.,

11) $2A \longrightarrow 20$, or   12) $3A \longrightarrow 31$, etc.

(b) The variable cycle-length, fixed sequence case:

$$G_{2B} = \langle V_T, V_N, P, S \rangle$$

where $V_T = \{0,1,2,3\}$, $V_N = \{A,B,C,S\}$,

and P consists of:

1) $S \longrightarrow A$	5) $B \longrightarrow BC$	8) $A \longrightarrow 0$	
2) $A \longrightarrow AA$	6) $C \longrightarrow CC$	9) $B \longrightarrow 1$	
3) $A \longrightarrow AB$	7) $C \longrightarrow CA$	10) $C \longrightarrow 2$	
4) $B \longrightarrow BB$		11) $2C \longrightarrow 23$	

Sample sentences in the language generated by G_{2B} include:

1)   0000001120111112223011230123
2)   012220112200011112223012
3)   012011230120000

Notice that one can have an arbitrary number of 0's followed by an arbitrary number of 1's, then an arbitrary number of 2's. The only restrictions are that (a) the first 1 or 2 of a string must be preceded by a 0 or 1, respectively; (b) if a 3 occurs, it must follow a 2 and only one 3 can occur at a time; and (c) the first 0 of a string must be preceded by a 2 or 3, unless it is the first symbol in a string.

C. Automata: Deciding Whether a String Belongs to a Language

One question that soon arises in considering grammars is deciding whether or not a given string of terminal symbols belongs to the language generated by a particular grammar. What is desired is a machine that tests for this.

Such machines exist for all the languages of interest in this paper and are referred to as sequential machines, or automata. In fact, a language can be characterized by an automaton which recognizes only sentences in that language by a grammar. I have chosen to follow grammatical models to describe the languages and to use automaton-like models for the animal acting (or recognizing) that language because it is conceptually clearer and because it leads to greater ease of relating structure to function in the animal.

Several types of automata exist. The simplest is probably the <u>finite-state</u> automaton. A finite-state automaton is a structure

$$A = \langle X, Q, Y, \delta, \lambda, q_0, F \rangle$$

where X is the set of terminal symbols encountered in any
string to be fed to it;

Q is the set of possible internal conditions or <u>states</u>, descriptive of the automaton itself;

Y is the set of symbols (or acts, etc.) emitted by the automaton;

δ is a function which specifies what the state of the automaton will be at the next point in time given the current state and current input. In symbols, $q(t+\triangle t) = \delta(x(t), q(t))$ where $x(t)$ is the input at time t and $q(t)$ is the automaton's state at time t;

λ is a function which specifies the symbol or act, $y \varepsilon Y$, emitted by the automaton as a function of its internal state. In symbols, $y(t) = \lambda(x(t))$ where $y(t)$ is the behavior emitted at time t;

q_0 is the state in which the automaton is at the beginning of a sentence; and

F is the state or set of states defined as final states. A string of symbols x_1, x_2 --- x_n is said to be <u>accepted</u> (that is, the string is a sentence in the language) if the automaton is in one of the states of F after the last symbol in the input string has been read.

The following automaton will, for example, accept the language generated by G_1 (see Ex.1 above):

$$A_1 = \langle X, Q, Y, \delta, \lambda, q_0, F \rangle$$

where $X = \{0, 1\}$, $Q = \{q_0, q_A, q_B, q_{ERR}\}$, $Y = \{0,1\}$,

$\delta(x,q)$ is defined so that:

$$\delta(0,q_0) = q_A \qquad \delta(1,q_A) = q_0 \qquad \delta(0,q_{ERR}) = q_{ERR}$$

$$\delta(1,q_0) = q_B \qquad \delta(0,q_B) = q_0 \qquad \delta(1,q_{ERR}) = q_{ERR},$$

$$\delta(0,q_A) = q_{ERR} \qquad \delta(1,q_B) = q_{ERR}$$

λ is defined such that:

$$\lambda(q_0) = 0 \qquad \lambda(q_B) = 0$$

$$\lambda(q_A) = 0 \qquad \lambda(q_{ERR}) = 1,$$

q_0 is q_0, and $F = \{q_0\}$

This may also be shown diagrammatically if we let circles represent states and we let arrows going from one state to another represent transitions. Along with each transition we show the associated input; the output associated with the state is shown inside the circle for the state in parentheses. Thus for A_1 we have the diagram shown in Fig. 1. Although their algebraic descriptions will not be given here, state-transition diagrams are given in Figs. 2 and 3 for automata accepting the languages generated by grammars G_{2A} and G_{2B}, respectively (see Ex. 2 above). In all three cases, the outside observer knows a string is rejected as soon as the automaton emits a "1".

Note that there is nothing unique about either the grammar generating a particular language nor about the automaton accepting it. Most languages are capable of being generated by many different grammars and are accepted by many automata. With this in mind we turn momentarily to the classification of languages and their automata.

Basic Types of Grammars and Automata

Chomsky (1959) originally categorized phrase-structure grammars in terms of the types of rules they incorporated. A type 0 language has no restrictions on rules: any string of symbols can be replaced by any other. If we make the restriction that the number of symbols in the string cannot exceed the number of symbols in the string replacing it, the resultant grammar is a type 1 or context-sensitive grammar. Grammar G_1 is of this type.

If the number of symbols on the left in a rule is restric-

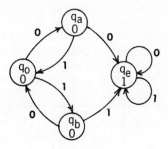

Fig. 1.   State-transition diagram for Automaton A_1.

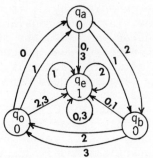

Fig. 2.   State-transition diagram for Automaton A_2 accepting the language generated by G_{2A}.

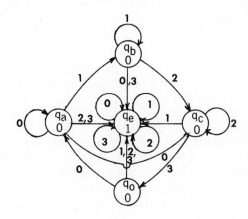

Fig. 3.   State-transition diagram for Automaton A_3 accepting the language generated by G_{2B}.

ted to one and the string on the right has at least one symbol, the grammar is said to by <u>type 2</u>, or <u>context-free</u>.
Finally, if the productions are all of either the form
A——>aB or A——>a where A and B are nonterminal symbols and
a is a terminal symbol, then the grammar is a <u>type 3</u> or a
<u>regular</u>, or a <u>finite-state</u> grammar. Note that every regular
grammar is context-free, every context-free grammar is context-sensitive and every context-sensitive grammar is type 0.
(I apologize for the confusing terminology but it's not mine.)

A language is said to be type 0 if there is a type 0
grammar that generates it but no context-sensitive grammar
that does. Context-free and regular languages are similarly
defined. The fact that there is a context-sensitive grammar
generating a language does not make the language context-
sensitive; it may be context-free or regular instead. For
example, the language generated by G_1 is regular even though
G_1 is not context-free, let alone regular.

Intuitively, regular languages are much simpler than
type-0 languages with context-free and context-sensitive
languages falling in between. From this, one might expect
that more complicated languages need more complicated
automata to "understand" them. Such is the case: context-
free, context-sensitive and type-0 languages require automata
known as pushdown automata, linear bounded automata and turing
machines to accept them, respectively. Because of this, most
of the work done on practical models has been done with regular and context-free languages; relatively little has been
done with context-sensitive models. More will be said on this
later when programmed grammars are discussed.

III.   Towards a Linguistic Model of an Animal in its
        Environment

In the previous section I summarized the basic properties
and types of phrase-structure grammars, the languages they
generate and the automata accepting them. In this section I
shall broaden the model of the automaton to approximate the
logical structure of a real animal. Then I shall discuss a
number of generalizations to phrase-structure grammars; the
generalizations discussed will enable us to account for those
factors discussed in Section I which are not included in other
models.

A. The Logical Structure of the Animal

To construct a model of an animal we need to take a number of things into consideration. For one, we need to

characterize the animal's <u>internal state</u>, i.e., the values of
the relevant variables describing its interior environment.
The internal state must include not only physiological information, such as lipid concentrations, internal temperatures
and electrolyte concentrations, but, in addition, it must include such memory and knowledge of the environment as the
animal possesses.

Along with this we need to understand how changes in internal state occur in response to external events. Hence we
need to include in our model both a set of internal states
(designated by Q) and some function describing state transitions as a function of current state, q, and current environmental state. Letting Ω denote the set of possible environmental states, the state transition function can be written
as $\delta(q,w)$, where the next state q_{i+1} is given by the relation

$$q_{i+1} = \delta(q_i, X_i), \quad q\epsilon Q, \ w\epsilon\Omega.$$

We also need to include in our model a description of the
behavior of the animal. Letting Y denote the set of behavioral acts the animal may engage in, it will be assumed that the
specific behavioral act chosen at any time is completely
determined by the animal's current state (remembering that the
animal's knowledge of the current environment is part of that
state). The function relating behavioral acts to internal
states will be denoted by λ, and the current behavior is given
by the relation

$$Yi = \lambda(q_i).$$

So far we have been following the path leading up to the
the finite state automata described in the last section.
There is a complication, however. Unlike a computer, an
animal is confronted with an event stream that does not arrive
in the form of discrete strings of neatly labelled terminal
symbols. What the animal <u>does</u> get is a collection of sensory
stimuli representing environmental information as filtered
through highly selective perceptual mechanisms. The animal
must then integrate and classify this information prior to
incorporating it in its decision-making procedures (as represented by $\delta(q,w)$ and $\delta(q)$). In essence, then, the animal contains an analyzer which accepts input from the environment and
feeds outputs into the animal's decision-making processes.
These outputs represent the assumed state of the environment.

The analyzer itself must, of course, be described. I
will assume that the animal is equipped with a set of feature
detectors which, in concert, allow it to discern the presence
or absence of the members of a set of features, X. In turn,
these features can be put together in various ways to form a

set of basic percepts, Ω. The members of the set Ω will be referred to as <u>events</u> hereafter.

The environment which the animal inhabits possesses a distinct structure in space and time. This spatiotemporal structure serves to organize and, in turn, is organized by the events occurring in it. The basic underlying space-time context will be referred to as a <u>quantal base</u>, $\mathcal{Q}$, and it is the means by which we shall measure the "shapes" and durations of events.

This leads us finally to our general working model of the logical structure of an animal. This structure will be described as a septuple

$$G_0 = \langle \mathcal{Q}, X, \Omega, Q, Y, \delta, \lambda \rangle$$

where $\mathcal{Q}, X, \Omega, Q, Y, \delta$ and λ are as defined above.

This definition does not take into account the fact that there may be syntactical restrictions as to which elements of Ω may follow which. The restrictions cannot properly be included because the sequence of environmental states is not completely under the animal's control. The question of "illegal" sequences (i.e., "unacceptable" ones) will be dealt with later.

Furthermore, the structure described is clearly a reasonable representation of the logical structure of the animal at only one level of description. In particular, it does not explicitly incorporate any information about the morphology and functional anatomy of the animal, nor is it obvious how one would create a model of this complexity on the basis of empirical data.

A fairly general method for resolving these kinds of problems has been proposed by Zeigler (1976). He has developed a method of modelling time-invariant systems (i.e., systems whose structure doesn't change with time) which leads to a <u>hierarchy</u> of sets of specifications rather than just one set. The lowest levels of hierarchy consist of descriptions of what output sequences go with what input sequences and are thus more or less behavioral in their orientation. Successively higher levels incorporate more and more of the logical structure of the system; eventually the system is described as a network of simpler systems, each possessing their own structure. These higher levels would clearly pertain to and involve a great deal of the actual physical and physiological structure of the animal.

One level of this series of specifications is occupied by what Zeigler refers to as an <u>iterative specification</u>. An iterative specification can be written as a septuple

$$G_0' = \langle T', X', \Omega', Q', Y', \delta', \lambda' \rangle$$

where T' is the time base (i.e., a quantal base which is
    limited to solely the time dimension),
    X' is a set of possible environmental states,
    Ω' is the set of possible elementary sequences of
    environmental states, any long sequences of en-
    vironmental states can be expressed as a se-
    quence of elements of Ω', and
    $Q', Y', \delta', \lambda'$ are as defined above.

Two deviations from Zeigler's model have been made. He
defined iterative specifications so they apply only to time-
invariant systems. In addition, G_0' has been redefined so as
to allow multidimensional quantal bases.
    A number of observations need to be made next:
    1) The fact that Zeigler assumed time-invariance in his
systems creates a problem. He assumed that, because the system doesn't change structurally, any input segment could be
followed by any other. I suspect that few, if any, animals
see the world in such terms. Ponds do not usually appear in
mid-air and fish do not often appear on deserts. In short,
there is syntax in the environment.
    2) Shifting our point of view for a moment to consider
the animal and its environment as a system of interest, it is
evident that the set of possible sequences of environmental
states form a language. Hence I argue that there is an
organism-specific language whose terminal symbols comprise the
animal's set Ω and whose syntax is parallel to that of the
environment. Thus language is effectively the animal's model
of the world, and will be denoted as L_ε.
    3) The set of sequences of internal states that would be
experienced by the animal also forms a language, as does the
set of sequences of behavior; these will be denoted as L_Q and
L_Y, respectively.
    4) There is a total behavioral language in which the
"acts" are pairs (q,y); these acts define the organism's total
response to environmental statements. Letting $v = (q,y)$ and
V be the set of all v, there is a language of total behavior
L_V whose terminals are V and whose productions are contingent
on inputs. More will be said on this later.
    5) On the basis of these remarks, the animal can be con-
sidered as a system which is engaged in an operational
dialogue with its environment; its "utterances" are its responses to its environment and are designed to mediate the interaction between its own changing needs and goals with the un-
folding monologue of its environment.
    These last remarks can be summarized in an alternate
description of the animal:

$$G_0'' = \langle L_\varepsilon, L_v, \Theta \rangle$$

where L_ε and L_v are as defined above, and Θ is the organism "translating" L_ε into L_v. Note that, in terms of our principle model, L_v is comprised of L_Q and L_γ, L_q is specified by the triple $\langle \Omega, Q, \delta \rangle$ and L_γ is specified by $\langle Q, Y, \lambda \rangle$.

B.   Extended Grammars for Behavioral Models

I showed in the last subsection that an animal can be modelled in terms of an input language, L_ε, and an output language, L_v. Let us look at these languages a little more closely. In particular, how applicable are the models introduced in Section II?

As they stand, those models are inadequate. Environmental languages generally require a multidimensional quantal base, but the traditional phrase-structure grammar generates only one-dimensional strings. It was argued in Section I that small variations in duration (and, I might add, spatial shape) may be very important semantically, yet the grammars considered in Section II only allowed for event durations of multiples of a common base unit. The traditional grammatical models assume no uncertainties in classification, yet the fact that the true set of environmental states is reduced by the animal's sensory systems to the set X argues that uncertainties in classification are bound to arise.

Thus we need to find ways to extend the models we considered in Section II to incorporate those factors. Four types of generalizations will be considered here: 1) generalizations of the forms of productions to allow for efficient incorporation of complex restrictions in grammars without unduly multiplying the number of rules or nonterminal symbols; 2) generalization of symbol shapes and durations to account for continuous variation in size or shape of symbol; 3) generalizations of rules defining membership of elements or sentences in a language to allow for uncertainties and ambiguities, and 4) generalizations to describe structures in more than one dimension, e.g., space and time.

Discussions of the automata accepting these languages have been omitted both because of limitations in space and because, in a few cases, not enough is known about them. See Fu (1974) for discussion of several types of automata accepting these more complex languages.

Generalizations of Forms of Productions

There is a certain unstructured quality about traditional phrase-structure grammars that arises from the lack of restrictions on the order in which the various rules are applied. It is tacitly assumed that any rule can be applied at any time that the string of symbols on which it operates appears. As a result, the languages generated tend to represent the art of the possible rather than the art of the reasonable. See Section II B for examples.

Such arbitrariness does not make good sense biologically. Hence it is tempting to include more and more of the surrounding context into a rule in order to delimit its range of applicability and thereby restrict the size of the language generated. This strategy rapidly leads to problems, since it will rapidly convert regular grammars into context-sensitive and even type-0 grammars, and we have seen that the more complicated grammars require more complicated automata to accept them. If we are to keep things manageable we should delimit ourselves to, at most, context-free languages.

Another approach is to modify the forms of rules by incorporating in each rule lists of permissible successors. This brings us to the notion of a programmed grammar.

Programmed grammars. Rosenkrantz (1970) has proposed a class of grammars, called programmed grammars, in which each rule is generalized to include two additional parts, one which defines the set of allowable successor rules if the production can be used in the current context (the success field), and the other defines the allowable successor rules if the present rule cannot be applied (the so-called failure field). For example:

Ex. 3 (see Fu, 1974). The "language" to be generated is the set of all equilateral triangles with sides of length 1, 2, or 3. This will be denoted as

$$L_3 = \{a^n b^n c^n \mid 1 \leq n \leq 3\}$$

where a, b, and c are line segments of unit length on each of the three sides of the triangle, respectively. Three grammars generating this language will be given. First, we show a programmed grammar:

$$G_{3A} = \langle V_T, V_N, P, S \rangle \text{ where } V_T = \{a,b,c\},$$

$V_N = \{S,B,C\}$, and P consists of:

Production #	Rule	Success	Failure
1	$S \longrightarrow aB$	$\{2,3\}$	$\{\emptyset\}$
2	$B \longrightarrow aBB$	$\{2,3\}$	$\{\emptyset\}$
3	$B \longrightarrow C$	$\{4\}$	$\{5\}$
4	$C \longrightarrow bC$	$\{3\}$	$\{\emptyset\}$
5	$C \longrightarrow C$	$\{5\}$	$\{\emptyset\}$

Note: Production "∅" means to halt, i.e., sentence done. A typical derivation would be as follows:

Step 1) S⟶aB (Rule 1). This is a forced choice, since Rule 1 is the only rule with S on the left-hand side. We choose our next rule from the list in the success field since we were able to execute Rule 1. We choose Rule 3.

Step 2) aB⟶aC (Rule 3). Since we were successful in applying Rule 3, we are forced to apply Rule 4 next since it is the only option in the success field.

Step 3) aC⟶abC (Rule 4). Since we were successful we are again forced in our choice; this time it's back to Rule 3.

Step 4) abC⟶abC (Rule 4: NOT applicable). This time we failed. So we take the next rule to be applied from the failure field: the choice is Rule 5.

Step 5) abC⟶abc (Rule 5). We are now forced to try Rule 5 again (since we succeeded in applying the rule).

Step 6) abc⟶abc (Rule 5: NOT applicable). Again we failed; this time we halt, since Rule ∅ is the only one specified on the failure field.

Comparing programmed grammars to other grammars generating the same language. As was said before, the same language can be generated by many other grammars. The choice of grammar is not completely arbitrary, however; different grammatical forms have different advantages and different weaknesses. Compare the preceding grammar to two others which generate the same language but are of different forms. From this we shall be able to see further advantages of the programmed grammar.

First, we consider a context-free grammar without structure in the production set:

$$G_{3B} = \langle V_T, V_N, P, S \rangle$$

where $V_T = \{a,b,c\}$, $V_N = \{S, A_1, A_2, B_1, B_2, B_3, C\}$ and P consists of:

1) $S \longrightarrow aA_1C$ 4) $A_1 \longrightarrow aA_2C$ 7) $B_2 \longrightarrow bB_1$
2) $A_1 \longrightarrow b$ 5) $A_2 \longrightarrow aB_3C$ 8) $B_1 \longrightarrow b$
3) $A_1 \longrightarrow aB_2C$ 6) $B_3 \longrightarrow bB_2$ 9) $C \longrightarrow c$

Finally, we may use a finite-state grammar to generate the same language:

$$G_{3C} = \langle V_T, V_N, P, S \rangle$$

where $V_T = \{a,b,c\}$, $V_N = \{S,A_1,A_2,B_{10},B_{20},B_{30},B_{21},B_{31},B_{32},B_{31},$
$B_{32},C_1,C_2,C_3\}$ and P consists of:

1)	$S \longrightarrow aA_1$	6)	$B_{10} \longrightarrow bC_1$	11)	$B_{32} \longrightarrow bC_3$
2)	$S \longrightarrow aB_{10}$	7)	$B_{20} \longrightarrow bB_{21}$	12)	$C_1 \longrightarrow c$
3)	$A_1 \longrightarrow aA_2$	8)	$B_{21} \longrightarrow bC_2$	13)	$C_2 \longrightarrow cC_1$
4)	$A_1 \longrightarrow aB_{20}$	9)	$B_{30} \longrightarrow bB_{31}$	14)	$C_3 \longrightarrow cC_2$
5)	$A_2 \longrightarrow aB_{30}$	10)	$B_{31} \longrightarrow bB_{32}$		

Notice that as we simplify the _forms_ of the productions we are
required to use more and more non-terminal symbols and productions. This tradeoff will be discussed later. The class of
languages generated by programmed grammars is broader than the
class of context-free languages but less broad than the context-sensitive languages (Fu, 1974). This makes them a
highly useful generalization. Furthermore, they can be made
stochastic by assigning probabilities to the various entries
in each success and failure field.

   A Hypothetical Biological Example. The choice of rule in
a programmed grammar can also be contingent on other variables.
In our case the determination might be made on the basis of
environmental conditions and/or internal state. Thereby this
class of grammar would seem to offer a number of interesting
possibilities for causal modelling of behavior sequences.
   Suppose, for example, we wish to model the foraging path
of a foraging bee. At any given flower it can either find
pollen or not: this will be the set of rewards, denoted as
R. (R = 1 means pollen found, R = 0 means no pollen found. We
assume that, if the bee finds pollen, it collects it.) Assume
that the bee needs to collect pollen from 50 flowers before
returning. We assume for simplicity that the flowers are
spaced so that the bee can fly 1 meter to find the nearest
flower, that going from one side of a clump to the other means
flying 5 meters and that clump-to-clump distance is 20 meters.
As the animal leaves it can make turns of 0°, 45°, 90°, or
135° right or left relative to its original line of flight.
   We shall further assume that the presence or absence of
pollen in the last two flowers (including the one currently
occupied) determines the bee's angle of departure relative to
the line of flight along which it arrived and distance to be
flown as follows (where r_x means a right turn of x degrees,
l_x means a left turn of x degrees, s = no turn, and the
probability of the turn follows in parentheses):

R (last flower)	R (this flower)	distance	angles(probability)
0	0	20	$r_{90}(.2), r_{45}(.2), s(.2),$ $l_{45}(.2), l_{90}(.2)$
0	1	1	$r_{90}(.3), r_{45}(.15), s(.1),$ $l_{45}(.15), l_{90}(.3)$
1	0	5	$r_{90}(.2), r_{45}(.2), s(.2),$ $l_{45}(.2), l_{90}(.2)$
1	1	1	$r_{135}(.3), r_{90}(.2), l_{90}(.2),$ $l_{135}(.3)$

When capacity is reached ($\Sigma R \geq 50$), the bee goes home. Likewise, if the number of failures, ΣF, is greater than 150 it goes home.

We may write this in the form of a stochastic programmed grammar:

Ex. 4:

$$G_4 = \langle V_T, V_N, P, B \rangle$$

where $V_T = \{r_{135}, r_{90}, r_{45}, s, l_{45}, l_{90}, l_{135}, d_1, d_5, d_{20}, h\}$

where d_x means fly x meters, and h means "go home",

$V_N = \{B, M_1, M_2, M_3, M_4\}$, and P is the following set of productions:

1)	$B \rightarrow M_1 M_2$	$\{2(.4), 3(.3), 4(.3)\}$	$\{\emptyset\}$
2)	$M_1 \rightarrow r_{45} d_{20}$	$\{5(R=1), 10(R=0)\}$	$\{\emptyset\}$
3)	$M_1 \rightarrow s d_{20}$	$\{5(R=1), 10(R=0)\}$	$\{\emptyset\}$
4)	$M_1 \rightarrow l_{45} d_{20}$	$\{5(R=1), 10(R=0)\}$	$\{\emptyset\}$
5)	$M_2 \rightarrow M_2 M_3$	$\{6(.25), 7(.25), 8(.25), 9(.25)\}$	$\{\emptyset\}$
6)	$M_2 \rightarrow r_{90} d_1$	$\{14(R=1), 19(R=0)\}$	$\{\emptyset\}$
7)	$M_2 \rightarrow r_{45} d_1$	$\{14(R=1), 19(R=0)\}$	$\{\emptyset\}$
8)	$M_2 \rightarrow l_{45} d_1$	$\{14(R=1), 19(R=0)\}$	$\{\emptyset\}$
9)	$M_2 \rightarrow l_{90} d_1$	$\{14(R=1), 19(R=0)\}$	$\{\emptyset\}$
10)	$M_2 \rightarrow M_2 M_3$	$\{11(.4), 12(.3), 13(.3)\}$	$\{\emptyset\}$
11)	$M_2 \rightarrow r_{45} d_5$	$\{25(R=1), 31(R=0)\}$	$\{\emptyset\}$
12)	$M_2 \rightarrow s d_5$	$\{25(R=1), 31(R=0)\}$	$\{\emptyset\}$

13) $M_2 \rightarrow l_{45} d_5$ $\{25(R=1), 31(R=0)\}$ $\{\emptyset\}$

14) $M_3 \rightarrow M_4 M_3$ $\{37$ if $\Sigma R > 50;\ 15(.3), 16(.2),$ $\{\emptyset\}$
                        $17(.2), 18(.3)\}$

15) $M_4 \rightarrow r_{135} d_1$ $\{14(R=1), 19(R=0)\}$ $\{\emptyset\}$

16) $M_4 \rightarrow r_{90} d_1$ $\{14(R=1), 19(R=0)\}$ $\{\emptyset\}$

17) $M_4 \rightarrow l_{90} d_1$ $\{14(R=1), 19(R=0)\}$ $\{\emptyset\}$

18) $M_4 \rightarrow l_{135} d_1$ $\{14(R=1), 19(R=0)\}$ $\{\emptyset\}$

19) $M_3 \rightarrow M_3 M_4$ $\{37$ if $\Sigma R > 150;\ 20(.2), 21(.2),$ $\{\emptyset\}$
                        $22(.2), 23(.2), 24(.2)\}$

20) $M_3 \rightarrow r_{90} d_5$ $\{25(R=1), 31(R=0)\}$ $\{\emptyset\}$

21) $M_3 \rightarrow r_{45} d_5$ $\{25(R=1), 31(R=0)\}$ $\{\emptyset\}$

22) $M_3 \rightarrow s d_5$ $\{25(R=1), 31(R=0)\}$ $\{\emptyset\}$

23) $M_3 \rightarrow l_{45} d_5$ $\{25(R=1), 31(R=0)\}$ $\{\emptyset\}$

24) $M_3 \rightarrow l_{90} d_5$ $\{25(R=1), 31(R=0)\}$ $\{\emptyset\}$

25) $M_3 \rightarrow M_3 M_4$ $\{37$ if $\Sigma R > 50;\ 26(.3), 27(.15),$ $\{\emptyset\}$
                        $28(.1), 29(.15), 30(.3)\}$

26) $M_3 \rightarrow r_{90} d_1$ $\{14(R=1), 19(R=0)\}$ $\{\emptyset\}$

27) $M_3 \rightarrow r_{45} d_1$ $\{14(R=1), 19(R=0)\}$ $\{\emptyset\}$

28) $M_3 \rightarrow s d_1$ $\{14(R=1), 19(R=0)\}$ $\{\emptyset\}$

29) $M_3 \rightarrow l_{45} d_1$ $\{14(R=1), 19(R=0)\}$ $\{\emptyset\}$

30) $M_3 \rightarrow l_{90} d_1$ $\{14(R=1), 19(R=0)\}$ $\{\emptyset\}$

31) $M_3 \rightarrow M_3 M_4$ $\{37$ if $\Sigma F > 150;\ 32(.2), 33(.2),$ $\{\emptyset\}$
                        $34(.2), 35(.2), 36(.2)\}$

32) $M_3 \rightarrow r_{90} d_{20}$ $\{25(R=1), 31(R=0)\}$ $\{\emptyset\}$

33) $M_3 \rightarrow r_{45} d_{20}$ $\{25(R=1), 31(R=0)\}$ $\{\emptyset\}$

34) $M_3 \rightarrow s d_{20}$ $\{25(R=1), 31(R=0)\}$ $\{\emptyset\}$

35) $M_3 \rightarrow l_{45} d_{20}$ $\{25(R=1), 31(R=0)\}$ $\{\emptyset\}$

36) $M_3 \rightarrow l_{90} d_{20}$ $\{25(R=1), 31(R=0)\}$ $\{\emptyset\}$

37) $M_4 \rightarrow h$ $\{\emptyset\}$ $\{\emptyset\}$

This grammar does not account for the type of serial correlation in angles reported by some investigators (Smith, 1975a,b) but the grammar could be enlarged to include such effects.

An Alternate Approach: Indexed Grammars. Another method for incorporating restrictions on rules has been suggested by Aho (1967). He has shown how a set of generalized grammars, called <u>indexed grammars</u>, can be created by introducing a finite set of flags to the grammar. The non-terminals in any string (or the right hand side of a production) can be followed by an arbitrary list of flags. If a terminal with flags is replaced by one or more non-terminals, the flags following each non-terminal are generated. If a non-terminal is replaced by a terminal the flags disappear. The flags then generate substrings of their own which depend on the non-terminals with which they are associated. For example (after Fu, 1974),

<u>Ex. 5</u>: $G_5 = \langle V_T, V_N, V_F, P, S \rangle$

where $V_T = \{0,1\}$, $V_N = \{S,T,A,B\}$, $V_F =$ the set of flags $\{f,g\}$, and P consists of:

1) $S \longrightarrow Tg$ 4) $f = \{A \rightarrow 0A, B \rightarrow 1B\}$
2) $T \longrightarrow Tf$ 5) $g = \{A \rightarrow 0, B \rightarrow 1 \}$
3) $T \longrightarrow ABA$

Applying Rule 1 once then Rule 2 (n-1) times and finally Rule 3, we get:

$$S \rightarrow Tg \rightarrow Tfg \rightarrow \text{---} \rightarrow Tf^{n-1}g \rightarrow Af^{n-1}g \ Bf^{n-1}g \ Af^{n-1}g.$$

Then using Rules 4 and 5,

$$Af^{n-1}g \ Bf^{n-1}g \ Af^{n-1}g \longrightarrow 0Af^{n-2}g \ 1Bf^{n-2}g \ 0Af^{n-2}g \longrightarrow \text{---}$$
$$\longrightarrow 0^{n-1}Ag1^{n-1}Bg \ 0^{n-1}Ag \longrightarrow 0^n1^n0^n.$$

That is, the language generated by G_5 is the language $0^n1^n0^n$, which is known to be context-sensitive. Thus, the set of languages generated by indexed-grammars includes some properly context-sensitive languages; it also includes the context-free languages.

Grammars with Non-Uniformly-Sized Elements

The problem of variable internal sizes was pointed out in Section I. In essence the problem lies in the fact that duration or shape may convey significant information <u>not</u> specifically carried by other features (Fentress, 1976a; Schleidt, 1973). Thus speed of execution and degree of stereotypy of movements may be expressive of arousal state (Fentress, 1976a). Similarly, an increase in running speed may be accompanied by a change in gait, as with horses.

If a class of terminal symbols has its own characteristic "size" and "shape", i.e., if any given type of symbol comes in only one size and shape, then no special account need be taken of spatial or temporal irregularity except insofar as might be required to fill the space in which the pattern is embedded. If, on the other hand, no relationship exists between shape and size vis-a-vis class of terminal symbol, then variability in size and shape can essentially be ignored. One approach to intermediate cases may lie in semi-Markov models but, as pointed out in Section I, this does not seem to have been done.

Non-Contextual Approaches. Two general approaches to considering non-uniformity of interval sizes can be identified. They differ as to whether sizes are considered in relative or absolute terms, i.e., whether context is taken into account. We shall consider non-contextual approaches first.

Deschamps (1974) has discussed the relations between synchronous and asynchronous languages and automata. Asynchronous languages are languages in which durations are all integer multiples of a common minimal duration, and are reducible to asynchronous languages by collapsing all strings of consecutive identical symbols to one occurrence of that symbol. Ex. 2B is an example of a language that can be so reduced.

Zeigler (1976) has approached the problem through the use of what he has termed a discrete event system specification (abbreviated DEVS). A DEVS is, formally speaking, a sextuple:

$$M = \left\langle X_m, S_m, Y_m, \delta_m, \lambda_m, t \right\rangle$$

where X_m is the set of external events;
S_m is the set of (sequential) states;
Y_m is the output value set (i.e., set of behaviors, etc.);
δ_m is the transition function;
λ_m is the output function; and
t_m is the time advance function.

t is a function of the state s; $t(s)$ is the maximum amount of time that the system can stay in state s. The state of the system is then defined by the sequential state s and the amount of time that the system has been in state s, denoted as e. That is, if q is the internal state, q = (s,e), $0 \le e \le t(s)$.

Suppose we define a special event not in X, called η. This "event" consists of nothing happening. The transition function is then defined in terms of s, e, and input events x (or η), as follows:

a) $\delta_m(s,e,\eta) = (s,e)$ so long as $e \leq \mathfrak{t}(s)$
b) $\delta_m^m(s,\mathfrak{t}(s),\eta) = (\delta_\eta(s),e)$ where $\bar{\delta}_\eta(s)$ is an
     autonomous state-transition function, and
c) $\delta_m(s,e,x \neq \eta) = (\delta_x(s),e)$ where x is equal to $\delta(x,s)$,
     i.e., the input-forced state-transition function.

In short, if no external events occur (i.e., η "happens")
the system stays in state s for a time $\mathfrak{t}(s)$, then transits to
the next state $\delta_\eta(s)$; however, if an event x occurs at time
$e < \mathfrak{t}(s)$, the system transits to the next state $\delta_m(s,e,x)$.
Whether e goes to o at transition time is open to specifica-
tion in the individual case. In addition, observe that λ is
a function over states (s,e), not just s.

This model has the distinct advantage that it can be
converted into an iterative specification of the form given
earlier (Zeigler, 1976). It would appear to be a promising
form for modelling the animal but less promising as a model
for the environment, unless inputs affecting the environmental
dynamics are known and can be modelled.

Contextual Approaches. In many cases the relative durations of movements as well as their sequence and rapidity are
critical to understanding their meaning. Such relational
information can be and is used to determine such things as
another person's mood. One can easily tell whether a person
is angry, pensive, distracted, etc., by observing him or her
doing a complex but well-learned task such as washing dishes.
At a more elementary level, the relative sizes and shapes of
different patch types may radically affect the biological
characteristics of a habitat in ways not simply dependent on
the absolute sizes of the patches. Similarly, the ability of
a wasp to find its nest may depend upon its ability to recognize the spatial relations of a set of landmarks and find the
site where the landmarks assume a certain visual relationship
to one another (Miller, et al., 1960).

Such relational structures are not well described by any
of the methods outlined above. What is needed is a somewhat
more holistic approach. We may begin by considering the
notion of a behavioral melody (a concept referred to by a
number of investigators, notably Fentress, 1976a; Gibson,
1950; Thorpe and Hall-Craggs, 1976; Volkelt, 1912). We shall
define a melody here as a series of distinct acts (notes)
with an associated characteristic structure of durations and
accents (rhythm). For purposes of the problem at hand our
interest lies in the rhythmic structure.

The characteristic rhythm of a sequence is the product
of three things: a) the relative durations of events,
b) the specific events associated with the intervals, and
c) the accents placed in the melody. Such accents can arise
from increases in relative intensity of events (stress

accents), increases in duration (agogic accents) or by changes
in melodic progression (e.g., changes in motivation or purpose
of behavior).

This formulation of the problem renders it generally
equivalent to the analysis of rhythm in music. Let us consider the type of methods in use there.

Cooper and Meyer (1960) have outlined a hierarchical
approach to the analysis of rhythm in music. The basic
elements at the lowest level are the feet, or groups of
accented and unaccented pulses, used in the analysis of
poetry: if "-" means accent and "u" means no accent, the
various kinds of feet are defined as follows:

1)   iamb: u -
2)   trochee: - u
3)   anapest: u u -
4)   dactyl: - u u
5)   amphibrach: u - u

By allowing beats to possess membership in more than one
foot (see subsection 3 below), Cooper and Meyer are able to
develop a useful and surprisingly rich analytical method.
Consider Ex. 4 illustrated in Fig. 4. While the obvious
rhythmic structure is a series of trochees (shown by the
downward-facing brackets), melodic coherency is maintained by
the iambs linking the trochees (indicated by the upward-
facing brackets). Note from Fig. 5 that shifting durations
of units changes the rhythmic structure so that the <u>iambs</u>

Fig. 4. *Lower level rhythmic analysis of a simple
melody (After Cooper and Meyer, 1960 with permission from
Univ. Chicago Press).*

Fig. 5. *Rhythmic analysis of a variant of "Twinkle,
Twinkle, Little Star".*

Fig. 6. *Rhythmic analysis of a fragment from Haydn's
"Surprise" Symphony (last movement).*

become the dominant rhythmic unit. A somewhat more complex example is shown in Fig. 6 (see Cooper and Meyer, 1960: pp. 65-88 for more detail). Similar methods have been proposed on this and other aspects of music by numerous authors (Berry, 1976; Salzer, 1952), while similar (albeit somewhat more speculative) analyses have been carried out in the visual arts (Burnham, 1971; Wollheim, 1968).

Grammars for Uncertain Situations

Animals constantly have to deal with uncertainty. Insofar as conventional grammars do not allow one to deal with uncertainty in a structured manner (see section III.B.), there is clear need to provide ways of dealing with such vagaries in linguistic models.

There are basically two types of uncertainty that can arise in linguistic models. The first is stochastic uncertainty. That is, uncertainty that arises because of the stochastic nature of the environment in which the events occur (or randomness in the animal's function). The other type of uncertainty is categorical uncertainty, that is, uncertainty as to whether a given event falls into zero, one, or more than one of the available classificational categories. The two cases will be discussed separately.

Grammars for Stochastic Uncertainties. A grammar can be rendered stochastic for purposes of modelling stochastic event streams by assigning to each production a probability of use. That is, when two or more productions are applicable, the one used is selected at random, and the probability of a given event being selected is given as part of the rule (an example of this was given in Ex. 4). See Arbib (1969), Fu (1974) and Paz (1871) for more comprehensive discussions of probabilistic grammars and automata.

Grammars for Categorical Uncertainties. Grammars have been developed to deal with situations in which there is categorical uncertainty. Such situations can arise in a number of ways which will be discussed later.

These grammars, called <u>fuzzy grammars</u>, are based on the notion of fuzzy sets as introduced by Zadeh (1965). In a system of fuzzy sets, each element X_i has a membership in each set X_j given by a membership function $\mu X_j(X_i)$, where μX_j can take on values between 0 and 1. (Compare this with traditional theory where μX_j is either 0 or 1, period.) The notion has been used to define fuzzy languages, fuzzy automata and fuzzy grammars. An ordinary grammar can be generalized to become fuzzy by associating with each production in P a weighting θ between 0 and 1, inclusive, so that a typical production is written as

$$P: \text{string } 1 \xrightarrow{\theta} \text{string } 2.$$

The fuzzy language generated from a fuzzy grammar has its membership function computed as follows: the membership of a given derivation (sequence of productions) is the smallest θ associated with that derivation. A given sentence may have a number of derivations; in any case the membership function is the largest value of the membership function of the various derivations (Thomasen, 1973).

Mizumoto, Toyoda and Tanaka (1972) were able to derive a large number of grammatical types, including various types of fuzzy grammars by introducing the notion of a pseudo-grammar. These authors have subsequently defined fuzzy grammars so that the grade of application of the rule to be used next is conditioned by the N rules used before in the derivation (Mizumoto et al., 1973). A similar result may be obtained by fuzzing up an appropriate programmed grammar.

Applications. Several types of situations lend themselves to treatment by these methods. Uncertainties in classification can be dealt with by assigning a given element to two or more classes and then working backwards to find which classification leads to the highest membership for the sequence in which the element is embedded. This, of course, assumes that the element belongs to only one class in the first place. This should be contrasted to the case where several things may be happening at the same time, hence multiple classification of a given element may be the appropriate thing to do. Recall also that in the previous section it was shown that rhythmic analyses often reveal that a single unitary element could belong in two or more categories (Figs. 4 and 5).

Multiple membership may arise in other ways. Higher order elements may overlap in space and/or time, using lower order elements in common. More particularly, an event that serves to terminate one sequence of behaviors may be part of the initiation of another sequence; this is referred to as behavior chaining (Hinde, 1970). We may account for this by means of a linked grammar. A linked grammar is one whose symbols have a bipartite structure. Each element has a center and a periphery; overlaps between elements involve only their peripheral parts. All composition rules are conditional upon the requirement that overlapping elements must agree in their respective areas of overlap. This allows us to incorporate a degree of semantic flexibility not otherwise attainable.

Fuzziness is also useful with respect to what might be called distributed meaning. The relevance of a cue (its ecological "meaning") lies not in itself but in the phenomena to which it points. Furthermore the ultimate consequences of a behavior may not be encountered for a considerable period of time (see Katz, 1974; Taylor, et al., 1974; Westman and Dunham, In Review, for the relevance of foraging behavior to life history profiles). This integrity of the whole is the basis of the concept of "melodies" in behavior (discussed above). The importance of such holistic or relational models at the level of perception and cognition is indicated by a number of authors (Arnheim, 1974; Givson, 1966; Gilbert, 1975; Meyer, 1956; Miller, Galanter and Pribram, 1960; Nelson, 1973; Thorpe and Hall-Craggs, 1976; Tolman, 1948). Finally, a number of investigators have constructed models attempting to incorporate some of these semantic concepts (e.g., Baird and Kelly, 1974; Hanson, Riseman and Fisher, 1976; Zadeh, 1975a,b,c).

Grammars in Several Dimensions

The representation of spatial structure in models can arise in three different ways, and it is important to distinguish them. First, the concern may be to model the spatial organization as it appears to a map-maker (the cartographic approach). Such descriptions have arisen most often in biology in connection with distributions of organisms in space (e.g., Pielou, 1969, 1975) or foraging strategies of animals (e.g., Kiester and Slatkin, 1974; Schoener, 1971).

Second, the concern may be to model the movements of an animal in an unfamiliar or changeable environment. The principle focus of most such models has been with strategies for prey/host searching or predator/parasite avoidance (Altmann, 1974; Emlen, 1973; Pulliam, 1974; Schoener, 1971; Treisman, 1975a,b). The grammatical model described in Ex. 4 above serves as a representative of this type of spatial model.

Thirdly, the concern may be to model the movements of an animal in an environment it knows well. While models of changeable or unfamiliar environments will tend to be stochastic and/or fuzzy, the models of a known environment should tend to be deterministic and non-fuzzy. Such methods tend to be more applicable to movements of animals in their home range or territory, or to trap-lining animals.

A Cartographic Model. We now consider the problem of creating a cartographic description. One approach is to simply include relations within the set of terminal symbols in order to produce a language whose sentences are a set of directions for building an environment. Consider the following example:

Ex. 7: We wish to create a language describing landscapes.
Assume that the landscapes contain neither hills nor rivers
for the sake of simplicity. We might use a grammar like the
following:

$$G_7 = \langle V_T, V_N, P, \langle\!\langle scene \rangle\!\rangle \rangle$$

where V_T = {E,N,S,W,$\odot$, $\eth$, φ, ψ,Λ,$\mathcal{C}$,uu, Δ, $\{$ }

E,N,S,W are operators, such that, for example, E(x,y)
        means "x is to the east of y"; E = east,
        N = north, S = south, W = west;
$\odot$(x,y) means "x is in y";
$\eth$(x,y) means "x is on top of y";
φ represents a tree;
ψ represents edible forbs;
Λ represents inedible forbs;
$\mathcal{C}$ represents a body of standing water;
Δ represents a rock;
uu represents bog plants; and
$\{$ represents a forest.
V_N ={$\langle$scene$\rangle$,$\langle$field$\rangle$,$\langle$forest$\rangle$,$\langle$pond$\rangle$,$\langle$swamp$\rangle$,$\langle$bog$\rangle$,
        $\langle$marsh$\rangle$,$\langle$rock$\rangle$,$\langle$OP$\rangle$}.

and P contains:

  1) $\langle$scene$\rangle \longrightarrow \langle$forest$\rangle$
  2) $\langle$scene$\rangle \longrightarrow \langle$field$\rangle$
  3) $\langle$forest$\rangle \longrightarrow \langle$OP$\rangle$($\langle$forest$\rangle$,$\langle$field$\rangle$)
  4) $\langle$field$\rangle \longrightarrow \langle$OP$\rangle$($\langle$field$\rangle$,$\langle$forest$\rangle$)
  5) $\langle$OP$\rangle \longrightarrow$ E
  6) $\langle$OP$\rangle \longrightarrow$ N
  7) $\langle$OP$\rangle \longrightarrow$ W
  8) $\langle$OP$\rangle \longrightarrow$ S
  9) $\langle$forest$\rangle \longrightarrow \{$
10) $\langle$forest$\rangle \longrightarrow \odot$($\langle$swamp$\rangle$,$\langle$forest$\rangle$)
11) $\langle$forest$\rangle \longrightarrow \odot$($\langle$pond$\rangle$,$\langle$forest$\rangle$)
12) $\langle$field$\rangle \longrightarrow \psi$
13) $\langle$field$\rangle \longrightarrow \langle$OP$\rangle$($\psi$,$\langle$field$\rangle$)
14) $\langle$field$\rangle \longrightarrow \Lambda$
15) $\langle$field$\rangle \longrightarrow \odot$($\langle$marsh$\rangle$,$\langle$field$\rangle$)
16) $\langle$field$\rangle \longrightarrow \odot$($\varphi$,$\langle$field$\rangle$)
17) $\langle$field$\rangle \longrightarrow \odot$($\langle$rock$\rangle$,$\langle$field$\rangle$)
18) $\langle$field$\rangle \longrightarrow \odot$($\langle$pond$\rangle$,$\langle$field$\rangle$)
19) $\langle$pond$\rangle \longrightarrow \mathcal{C}$
20) $\langle$swamp$\rangle \longrightarrow \eth$($\langle$forest$\rangle$,$\langle$pond$\rangle$)
21) $\langle$swamp$\rangle \longrightarrow \odot$($\langle$pond$\rangle$,$\langle$swamp$\rangle$)
22) $\langle$swamp$\rangle \longrightarrow \odot$($\langle$bog$\rangle$,$\langle$swamp$\rangle$)
23) $\langle$marsh$\rangle \longrightarrow \eth$($\langle$field$\rangle$,$\langle$pond$\rangle$)
24) $\langle$marsh$\rangle \longrightarrow \odot$($\langle$pond$\rangle$,$\langle$marsh$\rangle$)
25) $\langle$bog$\rangle \longrightarrow \eth$($\mathit{uu}$,$\langle$pond$\rangle$)

continued on next page

26) ⟨rock⟩ ⟶ ⟨OP⟩(⟨rock⟩,⟨rock⟩)
27) ⟨rock⟩ ⟶ ♂(△,⟨rock⟩)
28) ⟨rock⟩ ⟶ △

At each step the entire scene is described; each production serves to subdivide or redefine a portion of the area. Some sample descriptions are shown in Figs. 7 and 8.

Other Multidimensional Grammars. A number of other grammars have been proposed to describe patterns of dimension two or higher (Fu, 1974). These include plex grammars (Feder, 1971), web grammars (Pfaltz and Rosenfeld, 1969), string grammars (Shaw, 1969) and graph grammars (Pavlidis, 1972a,b). Ota (1975) has defined an interesting class of grammars, called <u>mosaic grammars</u>, which apply to planar pat-

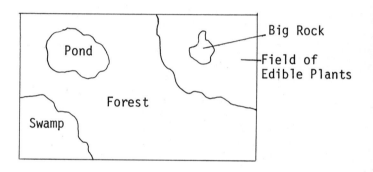

Fig. 7. Environment described by: $N(W(⊙(⌂,♀),$
$⊙(△,♀)), W(⊙(♂(♀,⌂), ♀), ♀)).$

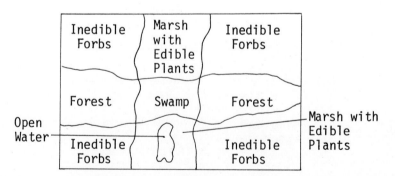

Fig. 8. Environment described by: $S(S(W(♪), W(⊙(⌂,♂$
$(♀,⌂)),♪)), W(♀, W(♂(♀,⌂), ♀))), W(♪, W(♂(♀,⌂),⊙$
$(△,♪)))).$

terns made of different-colored squares; these show promise
for describing patchy environments. I suspect that her
methods could be extended to patterns made of regular
hexagons.

One other class of languages applicable to cartographic
descriptions are the <u>cell-space languages</u>. Based on the notion of cellular automata (von Neumann, 1966) they have been
used to model self-reproduction (von Neumann, 1966; Thatcher,
1970), growth and morphogenesis (Howeweg and Hesper, 1974;
Laing, 1971) and to model the mammalian cerebellar cortex
(Mortimer, 1970). There is even a cell space computer
language implemented (Brender, 1970). Such languages have an
advantage insofar as they are framed to explicitly include
temporal variability. I should point out that there must be
clear relationships between the three types of spatial models
for a given animal in a given habitat. This has not been
explored, however, and forms a challenging area for future
research.

In closing this section I would like to deal with one
last problem. It was argued in Section I that one of the
deficiencies of traditional methods is the failure to incorporate higher level structure. While linguistic models are
essentially hierarchical, the hierarchical structure is
usually not explicitly stated. These relations can be brought
out if one notices the following facts (assuming levels are
numbered consecutively upward starting at 0 (for terminal
symbols):

1) A nonterminal symbol directly generating a terminal
symbol will be Level 1.

2) If $A \longrightarrow B$ in some production, then the level of A
is greater than or equal to that of B.

3) Suppose that there is some sequence of derivatives
such that, in essence, $A \longrightarrow B \longrightarrow C \longrightarrow --- \longrightarrow E \longrightarrow A$
Then all symbols in this loop must be at the same level.

The way this works can be illustrated from the following
example:

Ex. 6:

(a)   Let subscripts stand for level numbers, and the
original grammar is:

$$G_{6A} = \langle V_T, V_N, P, S \rangle$$

where V_T = {a,b,c}, V_N = {S,A,B,C}, and P consists of:

1) $S \longrightarrow A$ 4) $C \longrightarrow cA$ 6) $B \longrightarrow b$
2) $A \longrightarrow aB$ 5) $A \longrightarrow a$ 7) $C \longrightarrow c$
3) $B \longrightarrow bC$

then we may add subscripts according to the above rules and get:

$$G'_{6A} = \langle V'_T, V'_N, P', S_2 \rangle$$

where $V'_T = V_T$, $V'_N = \{S_2, A_1, B_1, C_1\}$, and P' consists of:

1') $S_2 \longrightarrow A_1$ 4') $C_1 \longrightarrow cA_1$ 6') $B_1 \longrightarrow b$
2') $A_1 \longrightarrow aB_1$ 5') $A_1 \longrightarrow a$ 7') $C_1 \longrightarrow c$
3') $B_1 \longrightarrow bc_1$

(b) If the original grammar is

$$G_{6B} = \langle V_T, V_N, P, S \rangle$$

where $V_T = \{a,b,c\}$, $V_N = \{S,A,B,C,D,E\}$, and P consists of:

1) $S \longrightarrow A$ 5) $B \longrightarrow bD$ 8) $C \longrightarrow c$
2) $S \longrightarrow aB$ 6) $C \longrightarrow cB$ 9) $D \longrightarrow dE$
3) $A \longrightarrow aC$ 7) $C \longrightarrow cC$ 10) $E \longrightarrow e$
4) $B \longrightarrow bC$

then we may add subscripts to get:

$$G'_{6B} = \langle V'_T, V'_N, P', S \rangle$$

where $V'_T = V_T$, $V'_N = \{S_5, A_4, B_3, C_3, D_2, E_1\}$, and P' consists of:

1') $S_5 \longrightarrow A_4$ 5') $B_3 \longrightarrow bD$ 9') $C_2 \longrightarrow C_1$
2') $S_5 \longrightarrow aB_3$ 6') $C_3 \longrightarrow cB_3$ 10') $C_1 \longrightarrow c$
3') $A_4 \longrightarrow aC_3$ 7') $C_3 \longrightarrow cC_3$ 11') $D_2 \longrightarrow eE_1$
4') $B_3 \longrightarrow bC_3$ 8') $C_3 \longrightarrow C_2$ 12') $E_1 \longrightarrow e$

(Rules 8',9', and 10' could be replaced by one rule, 8", where 8" is given by:

$$8") \; C_3 \longrightarrow cC_3.$$

Note that Rule 2' is also somewhat irregular.)

IV. IMPLEMENTING THE MODEL: BIOLOGICAL AND METHODS-LOGICAL CONSTRAINTS

We saw in the last section that syntactical models can be created incorporating the desiderata outlined in Section I. In fact a wide variety of syntactical structures can be erected to describe an animal. The problem is to find the appropriate one. To do so we need criteria beyond those outlined in previous sections. This, in turn, will allow us to deal with a question left unanswered: what does "not accepting" an environmental sentence mean in biological terms? To deal with these questions we need to place an evolutionary and ecological framework around our model.

A. Constructing an Evolutionary Model

For the moment envision the organism as a sequential decision-maker. At the end of each segment of its activity sequence it has to decide what its behavior will be during the next segment. It has as immediate sensory data its own internal state and its estimate of the environmental state. The behavior it chooses will obtain for the animal an immediate payoff which depends on its current state and that of the environment. Since its behavior will affect both its own internal state and, perhaps to a lesser degree, that of its surroundings; the animal's present behavior will affect its future payoffs either via internal state changes, changes in the environment, or both. This alteration in future payoffs, together with present payoffs, constitutes the total payoff for the animal's behavior.

Thus the animal has a prediction problem since the long-term payoffs for its behaviors may not be achieved (or even known) until some future time. Since the future payoffs will be contingent on future states of the environment, the animal has to act as if, to some degree, it is predicting the future states of the environment.

This payoff is measured in direct evolutionary terms: gain (or loss) in growth, survivorship and reproduction, viz., in fitness. The "goal" of the animal is to choose those behaviors which lead to the maximum evolutionary payoffs. There is, of course, a great big hitch. The energetic costs of maintaining the information-processing capabilities necessary to always choose the exact, best behavior is prohibitive --- assuming that it is even possible to predict that accurately. More particularly, it can be shown that, if the energetic costs of maintenance increase in some proportion to the informational channel capacity (Gallager, 1968) and benefits rise as the uncertainty reduction due to

information processing, natural selection will favor the
organism for which the marginal cost of increased complexity
is equal to the marginal expected payoff for such an increase
in complexity (Westman, In Review) (Fig. 9). The evolutionary
outcome of this process will be an organism which makes mis-
takes but which minimizes the evolutionary payoff reductions
due to its own errors.

Thus the operational language of an organism in its
environment is subject to two interacting constraints:
1) the complexity of the language represents a compromise
between inadequate benefits on the one hand and excessive
costs on the other, and 2) the syntactic and semantic structure of the organism's language should evolve so as to
minimize the costs of errors for the particular level of
complexity exhibited by the organism.

There is another constraint due to the physical structure
of the animal itself. Zeigler (1976) has shown how the type
of structural model we have been considering can itself be
broken down into an operational description in terms of the
physical subunits actually constituting the system. Hence the
general form of a specific animal should incorporate as much
of the functional organization and operating parameters of
the animal as is possible. This will also lead to a much more
easily tested model.

The foregoing remarks pertain more to the outlines of a
strategy for model-building and selection than to the tactics.
I should therefore like to consider four concerns to be dealt
with in constructing a model of a real animal: 1) how are
complexity and its costs to be measured? 2) how are the costs
and benefits of different strategies to be measured? 3) how
is error-cost minimization to be achieved? and 4) how are
these linguistic descriptions to be inferred from data?

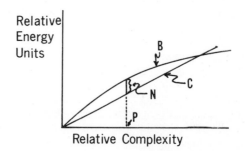

Fig. 9. Costs and benefits of different levels of com-
plexity. C = costs, B = benefits, N = net benefits = cost -
benefits; P is the point of maximum net benefits, hence is
the optimal complexity (after Westman, In Review).

Complexity and Its Costs

The measurement of complexity is rather problematic.
Consider again Ex. 3 (Section III). There we found that the
same language could be described by three grammars. G_{3A}
(the programmed grammar) used relatively few categorical
variables and productions, but it was necessary to use somewhat complicated types of productions to achieve this.
Grammar G_{3C} on the other hand used very simple productions but
it was necessary to introduce many different categorical
variables and productions to achieve this. Grammar G_{3B} is
clearly intermediate to the other two grammars.
      Which is the more complicated form? Let us see if we can
answer the question in biological terms. Assume that the
categorical variables correspond to discrete categories of
events or behaviors recognized by the organism and that the
productions represent neural circuits detecting or generating
relationships. Then G_{3A} would correspond to an organism which
recognizes relatively few categories of events and which
processes its information using a relatively small number of
complex neural centers. An organism embodying G_{3C} would
recognize a large number of categories (and is therefore
likely to have more sophisticated feature extraction
capabilities) and processes its information using a somewhat
larger number of simpler neural circuits. Because of the
particularly simple form of the productions, an organism
embodying G_{3C} might not possess any anatomically distinct
neural centers at all, whereas an organism embodying G_{3A}
would be much more likely to possess well defined nerve
centers, i.e., ganglia. It is interesting to note that the
process of cephalization tends to raise the effective complexity of available productions while reducing the number of
distinct neural centers; when this is accompanied by increasing sophistication in sensorimotor capabilities (i.e.,
the terminal symbols in the grammar) it is obvious that complexity is increasing since we can expect the resultant language to be more complex by almost any measure.

      Quantitative Measures of Complexity. As illuminating as
this approach may be it really doesn't leave us with a
useable quantitative measure of complexity; counting elements
in the grammar (whether they be terminal symbols, non-terminal
symbols, and/or productions) clearly won't suffice. One
approach is to try to determine complexity on the basis of
estimating the number of bits of information it would take to
construct the organism from building blocks. Using this
approach, Dancoff and Quasther made several estimates of the
information content of organisms (Dancoff and Quastler, 1953).
They obtained estimates between 5×10^{25} bits to 2×10^{28} bits

for an organism like an adult man. Apler (1966) has reviewed
this study as well as others by Raven (1961) and Elsasser
(1958) and found their estimates to be rather arbitrary and
the methods subject to serious criticism on biological
grounds.

Other approaches using information theory have been
developed that focus on the measurement of complexity in terms
of information processing rate rather than structural invest-
ment. Various investigators have derived entropy estimates
for probabilistic grammars (Justesen and Larsen, 1975; Rosen,
1974; Soule, 1974) and fuzzy grammars (Capocelli and DeLuca,
1973; DeLuca and Termini, 1972, 1974). On the basis of these
methods estimates of channel capacity (i.e., the maximum rate
at which information can be handled without errors) can and
have been derived (Soule, 1974). Such descriptions of complexity have the advantages that they lend themselves particularly well to studies of likely patterns of error-making
and ways to avoid specific errors.

Another point of view is a structural one, i.e., how complex is the automata required to understand the language?
We saw in Section II that the automata required to process
context-sensitive languages are more complex structurally than
those accepting context-free languages, and both are more
complex than finite-state automata. Williams (1975) has
pointed out that many of the higher dimensional grammars
require very complex automata to accept them. To what degree
an animal can lessen this demand by using the structure of its
surroundings to aid its memory (Langer, 1972; Miller et al.,
1960) is not clear. What is clear is that the biologically
relevant costs are those of growing and maintaining the
structure. The computation of structural maintenance costs
requires another approach.

Bremermann (1974) has reviewed several definitions of
complexity and argues strongly for the Rhodes definition of
complexity (Krohn and Rhodes, 1968; Rhodes, 1971). In crude
terms Rhodes complexity is a measure of the degree to which
the operation of a system is reversible, i.e., how much information is lost per unit operating time. Since information is
roughly equivalent to negative entropy, information loss is
entropy gain, and energy is required to offset entropy gain.

On the basis of this and Yost's (1973) finding that the
Rhode's complexity of an n-state neuron is n-1, Bremermann
derived a total heat dissipation for a human brain of 10^{-7}
watts, assuming that the human neuron is a 10-state device.
This is at first glance not a good estimate; von Neumann
has given an estimate of 10 watts (Bremermann, 1974). On the
other hand, Bremermann has argued that there are several
factors not taken into account which would tend to reduce the
discrepancy drastically.

In summary, information theoretic approaches to organismal complexity can yield flexible models for the comparison of different information-processing strategies, while measures based on the Krohn-Rhodes theory can be used in computing maintenance and operating costs. One other component of cost needs to be pointed out.

The development of complex organisms generally requires higher energy intakes, longer maturation times, and, possibly, more parental care. Hence certain components of the cost of complexity must be measured in ecological and life-historical terms. This leads to considerations of the costs and benefits of various life historical strategies.

Determination of Costs and Benefits of Different Strategies

This topic is one that has been of great concern to ecologists in recent years. While the concept of reproductive value has been around for some time (Fisher, 1958) only recently have investigators started to develop an analytical theory of life histories (c.f., Cole, 1954; Gadgil and Bossert, 1970; Schaffer, 1974; Taylor, et al., 1974). A parallel development has been taking place in the study of foraging and dietary strategies (c.f., Estabrook and Dunham, 1976; Pulliam, 1974; Schoener, 1971).

Most life history work has been done with essentially static models. Investigation of dynamic models (sensu Katz, 1974) has been attempted only a few times (Katz, 1974; Taylor, et al., 1974) although dynamical models have been employed in connection with predation and prey avoidance strategies (Anderson, 1976; Darwin and Williams, 1964; Treisman, 1975a,b).

A dynamical model of resource utilization strategies has been developed that is derived directly from the energy budgets of the animal (Westman and Dunham, In Review). This model predicts the net resource consumption and life history patterns in terms of foraging strategies and patterns of resource abundances. It can be used to determine total reproductive value as a function of age for different behavioral strategies and thus forms a complementary model to the ones proposed here.

As pointed out earlier, the animal's concern is to minimize the cost of its errors in decision-making. Thus, while this model provides a comprehensive scheme by which to compute costs and benefits of a given strategy, error cost minimization needs to be treated by other means.

Minimizing Error Costs

This problem exists at two levels. First, there is the

evolutionary level in which we are concerned with the evolution of what Conrad (1974a) has called the structural programs, viz., the evolution of the genotype. Second, there is
the problem of adaptively changing the parameters of that
structure so as to further reduce error costs, viz., the
development of the phenotype.

Studying either problem meaningfully presupposes that
one can restrict the search to a well-defined set of possible
models before a solution can be found. A number of methods
can then be brought to bear.

Holland has developed a very general and powerful method
for mathematically describing and studying adaptions in
complex systems (Holland, 1975). His methods are based on
the notion of an adaptive system. An adaptive plan consists
of a set of structures, a set of operators for modifying and
recombining elements of the structures, and a method for
determining what operators are to be applied to the structures
as a function of inputs. Included also is a method for
giving relative weights to the different structures as a
function of some criterion. The method has been applied
successfully to the creation of population genetics models and
to Britton and Davidson's (1969) model of gene regulation
(Holland, 1975). This approach is the best available for
modelling the evolution of the genotype (provided the class of
possible structures can be properly described). The approach
also is adapted to modelling learning in neurophysiological
terms; Holland has made a start, basing his work on Hebb's
(1949, 1958) framework. Pringle's (1975) notion of the central nervous system as a network of non-linear oscillators is
a viable and interesting alternative to the Hebbian basis, at
least in certain contexts.

A somewhat more restricted approach to the problem has
been outlined by Tsypkin (1971). He has outlined a series
of general models that exhibit adaptive and/or learning
characteristics. While his consideration is limited primarily
to man-made structurally programmed systems, he provides a
means of adaptively determining parameters in a manner that
may be more efficient than Holland's in certain contexts.

A third approach lies in the more traditional methods of
decision theory (see, for example, DeGroot, 1970; Raiffa,
1968). While some interesting studies of these methods have
appeared in the literature on pattern recognition (Ibaraki,
1976; Kelsey and Haddad, 1973; Tartara, 1973), few examples
have appeared in the biological literature (Dawkins and
Dawkins, 1973; Treisman, 1974a,b). Other interesting methods
involve algorithmic approaches to information theory in
decision making (Hiem, 1974; Tartara, 1973).

We have briefly considered how to account for the costs
and benefits of various strategies and how we might seek out

minimum cost-of-error strategies. To test these models we
need to compare them to data sets, i.e., we need to be able
to construct comparable forms on the basis of experimental
data.

*Grammatical Inference: Constructing Grammatical Models from
Empirical Data*

Grammatical inference is the process of determining which
of a given class of grammars best represents a given data set.
A number of methods have been proposed. Feldman (1967) has
described a method for inferring a deterministic finite state
grammar from a given data sample. Horning (1969, 1971) has
described a Bayesian inference algorithm to determine which of
a given class of stochastic grammars maximizes the
a posteriori conditional probability that a grammar was the
one in effect when the data sample was observed. Unfortunately, the method tends to be impractical if the number of
possible forms is too large due to excessive computation
costs (Fu, 1974). More complete descriptions of these and
other methods may be found in Fu (1974); see also Cook and
Rosenfeld (1976) and Wharton (1974).

Holland's (1975) methods also seem suited to this problem. However, no application of them has been made to
grammatical inference.

B. Accepting Environmental Languages

Early in Section III it was pointed out that the transition rules of an environmental language could not be incorporated into an animal because the animal never has complete
control over those transitions. Consequently there is a
question as to what happens when the organism encounters a
sequence it doesn't "accept". Obviously, the state function
$\delta(q,w)$ has to be defined for every pair (q,w): the animal
cannot just refuse to acknowledge that w has happened. We
shall deal with this problem in terms of payoffs.

Assume that we are concerned with a given sequence of
environmental states w_1w_2, ---, w_m, and that the organism's
initial state is q_o. Then in response to this sequence
of inputs the animal will generate a set of behaviors
y_1y_2, --- y_m and experience a set of internal states
q_1q_2, --- q_m. These behaviors will ultimately obtain a payoff
for the organism which is a function of the environmental
states and its own behavior. We may then argue that accepting
a given input sequence consists of emitting an appropriate
behavior sequence so as to obtain a total payoff exceeding
some threshhold.

More specifically, let $p = p(q_0, w_1 w_2 ---w_n)$ be the measure of long-term payoff for the system. Then, if $p_{min}(q_0)$ is the minimum payoff allowing survival for an animal in state q_0, we define the <u>acceptance function</u> as:

$$\mu(q_0, w_1 w_2 --- w_m) = \begin{cases} 0 \text{ if } \mu(q_0, w_1 w_2 --- w_m) < \mu_{min}(q_0), \\ 1 \text{ if } \mu(q_0, w_1 w_2 --- w_m) \geq \mu_{min}(q_0). \end{cases}$$

V. OVERVIEW

At the outset of this paper it was argued that the organism-environment interaction could be conceptualized as an operational dialogue and described in linguistic terms. After reviewing the basic attributes of grammars and automata, it was shown how they could be generalized in a number of ways to account for the effects of spatial structure, asynchronous spatiotemporal interval structure and uncertainties of classification and/or stochastic variability, and that these models could easily account for nonstationarity in the animal-environment system. Finally, it was argued that evolutionary bounds on complexity, the attendant criteria for minimization of cost of errors and the incorporation of known physical structure serve as restrictions to help delimit the class of possible models of the animal; the last section closed with a survey of some of the methods needed to practically implement the proposed methodology.

A. The Basic Model Revisited

The basic model of the organism that emerges from the preceding can be summarized as follows:
We assume the model animal lives in an environment E. This environment is conceptualized as a stream of events occurring in a space-time continuum described by the quantal base $\mathcal{Q}$. The animal gathers information about its environment through a sensory system that allows it to detect a certain set (x) of attributes of its environment. These attributes, or features, can be combined to form a set of events Ω which share delimited extents in space and time. Sequences of input states are generated using rules drawn from a set of productions Pe; hence the animal's functional model of its environment can be described as a language L_E, where L_E is generated by a grammar using Ω as its set of terminal symbols.
We further assume the organism's internal state is one of a set of possible internal states Q, and that the animal has

a set of behaviors Y. Changes in internal state are described by the function $\delta(q,w)$, where q is the current state and w is the current state of the environment. Similarly the behavior of the animal is given in terms of current internal state by the function $y = \lambda(q)$. Hence the total behavior of the animal can be described by a function T, where:

$$(q(t + \Delta t), y(t)) = T\ (q(t), x(t))$$
$$= (\delta(q(t), x(t)), \lambda(t)).$$

The set of paired values of q and y constitute the set of total behaviors (V), and the structure of the total behavioral repertoire can be written as another language L_V whose set of terminal symbols is V and whose productions P_V are determined as a programmed grammar by the inputs w(t) (see Ex. 4). Hence the animal "accepts" language L_E and "utters" sentences from L_V in return.

The language L_V can be broken down into an internal state language L_Q and an (external) behavioral language L_Y; L_Q is generated by a structural grammar whose terminal symbols are Q and whose productions P_Q are determined in part by the inputs w(t). Similarly L_Y is a transformed edition of L_Q simply because the current behavior is determined solely by the internal state q. If λ is given the more general form $\lambda = (q,w)$, L_Y becomes a structural grammar whose terminal symbols are the set Y and whose transitions are conditional by the inputs w(t). Formally:

$$G_o = \langle \mathcal{Q}, X, \Omega, P_e, Q, Y, \delta, \lambda \rangle$$

Because of uncertainties in categorization the language L_E should be fuzzy; clearly it will be probabilistic in most cases. It will also be linked because of distributive meaning, and it will in general be rhythmic in structure. The language L_V will be fuzzy, linked and rhythmical, as will the component sub-languages L_Q and L_Y. In specific cases one or more of these attributes may not be needed to model these languages.

B. Discussion

How well does this conceptualization correspond to the real world? At a more theoretical level, the type of automata-theoretic structure postulated here is consistent with that of numerous other investigators (Brindley, 1969; Findlay and Daniell, 1973; Muller and Taylor, 1973, 1976). Conrad (1974a,b) has argued that the general neural net is a structural program that is modified in its biochemical characteristics through experience. This notion draws a great deal of support from the great plasticity and functional versatility of the neuron

(Bullock, 1976; Kandel and Spencer, 1968; Wachtel and Kandel, 1967). Many neural and hormonal systems show the character-istics postulated here (Davis, et al., 1974a,b; Marler, 1976; Nottebuhm, 1970; Siegler, et al., 1974).

The qualitative structural aspects indicated here lead to explanations for observed patterns of behavior which are not explained otherwise. Deviations from "optimal" behavior can be explained in terms of the use of heuristic rules-of-thumb on the part of the animals. This would explain the failure of experimental animals to behave in accordance with simple optimality models, such as was found by Emlen and Emlen (1975) and Krebs et al.(1974). Similarly, individual differences in strategies (Slater, 1974; Will, 1974) can be explained in terms of variations in the underlying structures.

The general theory also accounts for the complex patterns of evolution seen in mimicry systems (Papageorgis, 1975; Wickler, 1968) and the behavior of predators in such systems (Estabrook and Jesperson, 1974). Furthermore, some of the more remarkable life history patterns where a number of considerations produce substantially different individual life histories can be satisfactorily analyzed in terms of the arguments in Section IV (Martin, 1974). In addition, more quantitative studies of regulatory behavior (Henwood, 1975; Huey and Slatlein, 1975) are completely consistent with this framework.

There is clear reason to believe that the current framework may prove useful in modelling and understanding behavior. But have we come any closer to crossing the great barriers that separates our understanding from those of our fellow species? We need to listen to them to hear the answer.

ACKNOWLEDGEMENTS

I wish foremost to indicate my profound debt to my colleague, friend, and wife, Alida S Westman, with whom I have discussed these ideas over and over again and whose in-sight and perspective has never been less than invaluable. I wish also to thank B.P. Ziegler and J.H. Holland, whose help and personal interest have been of great significance to me. Much valuable discussion and criticism has been con-tributed by A.E. Dunham, and I also am grateful to J. Joy and S. Tonsor for discussions while these ideas were in their formative stages. Last, but by no means least, I wish to express my appreciation and indebtedness to B. Hazlett who, as a mentor and editor, has been of enormous help in getting this paper into the English language.

REFERENCES

Aho, A.V. 1967. Indexed grammars: an extension of context-free grammars. IEEE Ann. Symp. Switching and Automata Theory, 8th, Austin, Texas, Conf. recora.

Alcock, J. 1973. Cues used in searching for food by red-winged blackbirds Agelaius phoenicius. Behaviour 46: 174-188.

Altmann, S.A. 1974. Baboons, space, time and energy. Amer. Zool. 14: 221-248.

Anderson, D.R. 1975. Optimal exploitation strategies for an animal population in a Markovian environment: a theory and an example. Ecology 56: 1281-1297.

Apter, M.J. 1966. Cybernetrics and Development. Pergamon Press, Oxford.

Arbib, M.A. 1969. Theories of Abstract Automata. Prentice-Hall, Englewood Cliffs, N.J.

Arnheim, R. 1974. Art and Visual Perception. (New Ed.), Univ. of California Press, Berkeley.

Baerends, G.P., R.H. Drent, P. Glas and H. Groenewold. 1970. An ethological analysis of incubation behavior in the herring gull. Behaviour Suppl. 17: 135-235.

Baird, M.L. and M.D. Kelley. 1974. A paradigm for semantic pattern recognition. Pattern Recognition 6: 61-74.

Berry, W. 1976. Structural Functions in Music. Prentice-Hall, Englewood Cliffs, N.J.

Bolles, R.C. 1960. Grooming behavior in the rat. J. Comp. Physiol. Psychol. 53: 306-310.

Boring, E.G. 1950. A History of Experimental Psychology. Appleton-Century-Crofts, New York.

Box, G.E.P. and G.M. Jenkins. 1976. Time-Series Analysis: Forecasting and Control. (Rev. Ed.) Holden-Day, San Francisco.

Bremermann, H. 1974. Complexity of automata, brains and behavior. (In) Mathematics and Physics of the Nervous System (M. Conrad, W. Guttinger and M. Dal Cin, eds.). Springer-Verlag, Berlin.

Brender, R.F. 1970. A Programming System for the Simulation of Cellular Spaces. Tech. Report 07449-25-T, Computer and Communications Sciences Dept., The Univ. of Michigan, Ann Arbor.

Brindley, G.S. 1969. Nerve net models of plausible size that perform many simple learning tasks. Proc. Roy. Soc. Lond. B 174: 173-191.

Britten, R.J. and E.H. Davidson. 1969. Gene regulation for higher cells: a theory. Science 165: 349-357.

Burt, M.K. 1971. From Deep to Surface Structure: An Intro-duction to Transformational Syntax. Harper & Row, N.Y.

Bullock, T.H. 1976. In search of principles of neural

integration. (In) Simpler Networks and Behavior (J.C. Fentress, ed.). Sinauer Assoc., Sunderland, MA.

Burnham, J. 1971. The Structure of Art. George Braziller, N.Y.

Campbell, D.J. and E. Shipp. 1974. Spectral analysis of cyclic behaviour with examples from the field cricket Teleogryllus commodus (Walk.). Anim. Behav. 22: 862-875.

Cane, V. 1959. Behavior sequences as semi-Markov chains. J. Roy. Stat. Soc. Ser. B 21: 36-58.

Capocelli, R.M. and A DeLuca. 1973. Fuzzy sets and decision theory. Information and Control 23: 446-473.

Chatfield, C. and R.E. Lemon. 1970. Analyzing sequences of behavioral events. J. Theor. Biol. 29: 427-445.

Chomsky, N. 1956. Three models for the description of language. IEEE Trans. Information Theory IT-2: 113-124.

_____. 1957. Syntactic Structures. Mouton and Co., The Hague.

_____. 1959. On certain formal properties of grammars. Information and Control 2: 137-167.

_____. 1965. Aspects of the Theory of Syntax. M.I.T. Press, Cambridge.

Cole, L.C. 1954. The population consequences of life history phenomena. Quart. Rev. Biol. 29: 103-137.

Conrad, M. 1974. Molecular automata. (In) Mathematics and Physics of the Nervous System (M. Conrad, W. Guttinger and M. Dal Cin, eds.). Springer-Verlag, Berlin.

_____, W. Guttinger and M. Dal Cin (Eds.). 1974. Mathematics and Physics of the Nervous System. Springer-Verlag, Berlin.

Cook, C.M., Rosenfeld, A. and A.R. Aronson. 1976. Grammatical inference by hill climbing. Information Sciences 10: 59-80.

Cooper, G. and L.B. Meyer. 1960. The Rhythmic Structure of Music. Univ. of Chicago Press, Chicago.

Dacey, M.F. 1963a. Two-dimensional random point patterns: a review and interpretation. Papers of the Regional Science Assoc. 11: 41-55.

_____. 1963b. Order neighbor statistics for a class of random patterns in multidimensional space. Annals Assoc. Amer. Geographers 53: 505-575.

Dacey, M. 1965. A Review on Measures of Contiguity for Two and k-Color Maps. Tech. Report 2 of ONR Task No. 389-140. Dept. of Geography, Northwestern Univ., Evanston, Ill.

_____. 1969. Some properties of a cluster point process. Canad. Geographer 13: 128-140.

Dancoff, S.M. and H. Quastler. 1953. The information content and error rate of living things. (In) Information Theory in Biology (H. Quastler, ed.). Univ. of Illinois Press, Urbana.

Darwin, J.H. and R.M. Williams. 1964. The effect of time of hunting on the size of a rabbit population. N. Z. J. Science 7: 341-352.

Davis, W.J., Mpitsos, G.J. and J.M. Pinneo. 1974a. The behavioral hierarchy of the mollusk Pleurobranchaea. I. The dominant position of the feeding behavior. J. Comp. Physiol. 90: 207-224.

_____, _____ and _____. 1974b. The behavioral hierarchy of the mollusk Pleurobranchaea. II. Hormonal suppression of feeding associated with egg-laying. J. Comp. Physiol. 90: 225-243.

Dawkins, R. 1976. Hierarchical organization: a candidate principle for ethology. (In) Growing Points in Ethology (P.P.G. Bateson and R.A. Hinde, eds.). Univ. of Cambridge Press, Cambridge.

_____ and M. Dawkins. 1973. Decisions and the uncertainty of behavior. Behaviour 45: 83-103.

DeGroot, M. 1970. Optimal Statistical Decisions. McGraw-Hill, N.Y.

Delius, J.D. 1969. A stochastic analysis of the maintenance behaviour of skylarks. Behaviour 33: 137-178.

DeLuca, A. and S. Termini. 1972. A definition of a non-probabilistic entropy in the setting of fuzzy sets. Information and Control 20: 301-312.

_____ and _____. 1974. Entropy of L-fuzzy sets. Information and Control 24: 55-73.

Deschamps, J.P. 1974. Asynchronous automata and asynchronous languages. Information and Control 24: 122-143.

Elsasser, W.M. 1958. The Physical Foundation of Biology. Pergamon Press, London.

Emlen, J.M. 1973. Ecology: An Evolutionary Approach. Addison-Wesley, Reading, M A.

_____ and M.G.R. Emlen. 1975. Optimal choice in diet: test of a hypothesis. Amer. Nat. 109: 427-435.

Estabrook, G.F. and A.E. Dunham. 1976. Optimal diet as a function of absolute abundance, relative abundance and relative value. Amer. Nat. 110: 401-413.

_____ and D.C. Jesperson. 1974. Strategy for a predator encountering a model-mimic system. Amer. Nat. 108: 443-457.

Fabricius, E. and A.-M. Jansson. 1963. Laboratory observations on the reproductive behavior of the pigeon (Columbia livia) during the pre-incubation phase of the breeding cycle. Anim. Behav. 11: 534-547.

Feder, J. 1971. Plex languages. Information Sciences 3: 225-241.

Feldman, J.A. 1967. First Thought on Grammatical Inference. Stanford Artificial Intelligence Proj. Memo No. 55. Stanford Univ., Stanford, C A.

Fentress, J.C.   1972. Development and patterning of movement
    sequences in inbred mice. (In) The Biology of Behavior
    (J.A. Kiger, ed). Oregon State Univ. Press, Corvallis.
_____ (Ed.). 1976a. Simpler Networks and Behavior. Sinauer
    Assoc., Sunderland, M A.
_____. 1976b. Behavioral networks and the simpler systems
    approach. (In) Simpler Networks and Behavior (J.C. Fen-
    tress, ed.). Sinauer Assoc, Sunderland, M A.
_____ and F.P. Stilwell. 1973. Grammar of a movement sequence
    in inbred mice. Nature (Lond.) 224: 52-53.
Findlay, J.M. and G.J. Daniell. 1973. A model for pattern
    recognition by cell networks. J. Theor. Biol. 38: 641-645.
Fisher, R.A. 1958. The genetical theory of natural selec-
    tion. Dover, N.Y.
Fu, K.S. 1974. Syntactic Methods in Pattern Recognition.
    Academic Press, N.Y.
_____ and B.K. Bhargava. 1973. Tree systems for syntactic
    pattern systems. IEEE Trans. Comput. C-22: 1087-1099.
Gadgil, M. and W.H. Bossert. 1970. Life historical conse-
    quences of natural selection. Amer. Nat. 104: 1-24.
Gallager, R.G. 1968. Information Theory and Reliable
    Communication. J. Wiley and Sons, N.Y.
Garcia, J., Clarke, J.C. and W.G. Hankins. 1973. Natural
    responses to scheduled rewards. (In) Perspectives in
    Ethology (P.P.G. Bateson and P.H. Klopfer, eds.). Plenum
    Press, N.Y.
Gibson, J.J. 1966. The Senses Considered as Perceptual
    Systems. Houghton-Miflin, Boston.
Gilbert, L.E. 1975. Ecological consequences of a coevolved
    mutualism between butterflies and plants. (In) Coevolution
    of Animals and Plants (L.E. Gilbert and P.H. Raven, eds).
    Univ. of Texas Press, Austin.
Gordon, M.S., Bartholomew, G.A., Grinnell, A.D., Jorgensen, C.
    B., and F.N. White. 1972. Animal Physiology: Principles
    and Adaptations. Macmillan, N.Y.
Goss-Custard, J.D. 1970. The response of redshank (Tringa
    totanus (L)) to spatial variations in their prey density.
    J. Anim. Ecol. 39: 91-113.
Hanson, A.R., Riseman, E.M. and E. Fisher. 1976. Context in
    word recognition. Pattern Recognition 8: 35-45.
Hazlett, B.A. and G.F. Estabrook. 1974a. Examination of
    agonistic behavior by character analysis. I. The spider
    crab Microphrys bicornutus. Behaviour 48: 131-144.
_____ and _____. 1974b. Examination of agonistic behavior by
    character analysis. II. Hermit crabs. Behaviour 49:
    88-110.
Hebb, D.O. 1949. The Organization of Behavior. J. Wiley and
    Sons, N.Y.
_____. 1958. A Textbook of Psychology. Saunders,

Philadelphia.

Heiligenberg, W. 1973. Random processes describing the
occurrence of behavioral patterns in a cichlid fish.
Anim. Behav. 21: 169-182.

_____. 1976. A probabilistic approach to the motivation of
behavior. (In) Simpler Networks and Behavior (J.C. Fentress, ed.). Sinauer Assoc., Sunderland, M A.

Heim, R. 1974. An algorithmic approach to information
theory. (In) Mathematics and Physics of the Nervous System
(M. Conrad, W. Guttinger and M. Dal Cin, eds.). Springer-
Verlag, Berlin.

Henwood, K. 1975. A field-tested thermoregulation model for
two diurnal Namib desert Tenebrionid beetles. Ecology 56:
1329-1342.

Hinde, R.A. 1970. Animal Behavior: A Synthesis of Ethology
and Comparative Psychology. 2nd Ed. McGraw-Hill, N.Y.

Hockett, C.F. and S.A. Altmann. 1968. A note on design
features. (In) Animal Communication (T.A. Sebeok, ed.).
Indiana Univ. Press, Bloomington.

Hogeweg, P. and B. Hesper. 1974. A model study on biomorpho-
logical description. Pattern Recognition 6: 165-179.

Holland, J.H. 1975. Adaptation in Natural and Artificial
Systems. Univ. of Michigan Press, Ann Arbor.

Hopcroft, J.E. and J.D. Ullman. 1969. Formal Languages and
their Relation to Automata. Addison-Wesley, Reading, MA.

Horning, J.J. 1969. A Study of Grammatical Inference. Tech.
Rpt. No. CS-139, Comput. Sci. Dept., Stanford Univ.,
Stanford, C A.

_____. 1971. A procedure for grammatical inference. IFIP
Congr., Yugoslavia.

Hoyle, G. 1976. Approaches to understanding the neurophysio-
logical bases of behavior. (In) Simpler Networks and
Behavior (J.C. Fentress, ed.). Sinauer Assoc., Sunderland,
M A.

Huey, R.B. and M. Slatkin. 1976. Cost and benefits of lizard
thermoregulation. Quart. Rev. Biol. 51: 363-384.

Huijbregts, C.J. 1975. Regionalized variables and quantitative analysis of spatial data. (In) Display and Analysis
of Spatial Data (J.C. Davis and M.J. McCullagh, eds.).
J. Wiley and Sons, London.

Hutt, S.J. and C. Hutt. 1970. Direct Observation and
Measurement of Behavior. Thomas, Springfield, Ill.

Ibaraki, T. 1976. Finite automata having cost functions.
Information and Control 31: 153-176.

Johnson, L.K. and S.P. Hubbell. 1974. Aggression and competition among stingless bees: field studies. Ecology 55:
120-127.

Justesen, J. and K.J. Larsen. 1975. On probabilistic context-free grammars that achieve capacity. Information and

Control 29: 268-285.

Kalmus, H. 1969. Animal behaviour and theories of games and of language. Anim. Behav. 17: 607-617.

Kandel, E.R. and W.A. Spencer. 1968. Cellular neurophysiological approaches in the study of learning. Physiol. Rev. 48: 65-134.

Kashyap, R.L. and A.R. Rao. 1976. Dynamic Stochastic Models from Empirical Data. Academic Press, N.Y.

Katz, P.L. 1974. A long-term approach to foraging optimization. Amer. Nat. 108: 758-782.

Kelsey, D.W. and A.H. Haddad. 1973. Detection and prediction of a stochastic process having multiple hypotheses. Information Sciences 6: 301-311.

Kiester, A.R. and M. Slatkin. 1974. A strategy of movement and resource utilization. Theoret. Popul. Biol. 6: 1-20.

Klir, G.J. 1969. An Approach to General Systems Theory. Van Nostrand Reinhold, N.Y.

Knuth, D.E. 1968. Semantics of context-free languages. J. Math. Syst. Theory 2: 127-146.

Krebs, J.R. 1973. Behavioral aspects of predation. (In) Perspectives in Ethology (P.P.G. Bateson and P.H. Klopfer, eds.). Plenum Press, N.Y.

_____, J.C. Ryan and E. Charnov. 1974. Hunting by expectation or optimal foraging? A study of patch use by chicadees. Anim. Behav. 22: 953-964.

Krohn, K. and J.L. Rhodes. 1968. Complexity of finite semigroups. Annals of Math. 88: 128-160.

Laing, R.D. 1971. Formalisms for biology: a hierarchy of developmental processes. Intern. J. Neuroscience 2: 219-232.

Langer, S.K. 1972. Mind: An Essay on Human Feeling. Vol. II, Pt. 4: The Great Shift. Johns Hopkins Univ. Press, Baltimore.

Lemon, R.E. and C. Chatfield. 1971. Organization of song in cardinals. Anim. Behav. 19: 1-17.

Lenneberg, E.H. 1964. A biological perspective on language. (In) New Directions in the Study of Language (E.H. Lenneberg, ed.). M.I.T. Press, Cambridge.

Marler, P. 1976. Sensory templates in species-specific behavior. (In) Simpler Networks and Behavior (J.C. Fentress, ed.). Sinauer Assoc., Sunderland, M.A.

Martin, S.G. 1974. Adaptations for polygynous breeding in the Bobolink Dolichonyx oryzivorus. Amer. Zool. 14: 109-119.

Maurus, M. and H. Pruscha. 1973. Classification of social signals in squirrel monkeys by means of cluster analysis. Behaviour 47: 106-128.

May, R.M. 1973. Stability and Complexity in Model Ecosystems. Princeton Univ. Press, Princeton.

Meyer, L.B. 1956. Emotion and Meaning in Music. Univ. of
    Chicago Press, Chicago.
Miller, G.A., Galanter, E. and K.H. Pribram. 1960. Plans and
    the Structure of Behavior. Holt, Rhinehart and Winston,
    N.Y.
Mizumoto, M., Toyoda, J. and K. Tanaka. 1972. General for-
    mulation of formal grammars. Information Sciences 4:
    87-100.
_____, _____ and _____. 1973. N-fold fuzzy grammars.
    Information Sciences 5: 25-43.
Morgan, B.J.T., Simpson, M.J.A., Hanby, J.P. and J. Hall-
    Craggs. 1976. Visualizing interactions and sequential
    data in animal behavior: theory and application of
    cluster-analysis methods. Behaviour 56: 1-43.
Mortimer, J.A. 1970. A Cellular Model for Mammalian Cere-
    bellar Cortex. Tech. Report 03296-7-T, Computer and
    Communication Sciences Dept., The University of Michigan,
    Ann Arbor.
Muller, F.J. and W.K. Taylor. 1973. A comparative study of
    electronic and neural networks involved in pattern recog-
    nition. J. Theor. Biol. 41: 97-118.
_____ and _____. 1976. A parallel processing model for the
    study of possible pattern recognition mechanisms in the
    brain. Pattern Recognition 8: 47-52.
Murdoch, W.W. and A. Oaten. 1975. Predation and population
    stability. Advances in Ecol. Res. 9: 1-130.
Nelson, K. 1964. The temporal pattern of courtship behavior
    in the glandulocaudine fishes (Ostariophysi, Characidae).
    Behaviour 24: 90-146.
_____. 1973. Does the holistic study of behavior have a
    future? (In) Perspectives in Ethology (P.P.G. Bateson and
    P.H. Klopfer, eds.). Plenum Press, N.Y.
Nottebohm, F. 1970. Ontogeny of bird song. Science 167:
    950-956.
Ota, P.A. 1975. Mosaic grammars. Pattern Recognition 7:
    61-65.
Papageorgis, C. 1975. Mimicry in neotropical butterflies.
    Amer. Scientist 63: 522-532.
Pavlidis, T. 1972a. Graph theoretic analysis of pictures.
    (In) Graphic Languages (F. Nake and A. Rosenfeld, eds.).
    North Holland Publ., Amsterdam.
_____. 1972b. Linear and context-free graph grammars. J.
    Assoc. Comput. Mach. 19: 11-22.
Paz, A. 1971. Introduction to Probabilistic Automata.
    Academic Press, N.Y.
Pfaltz, J.L. and A. Rosenfeld. 1969. Web grammars. Proc.
    Joint Conf. Artificial Intelligence 1st, Washington, D.C.:
    609-619.
Pielou, E.C. 1969. An Introduction to Mathematical Ecology.

J. Wiley and Sons, N.Y.

_____. 1975. Ecological Diversity. J. Wiley and Sons, N.Y.

Pringle, J.W.S. 1951. On the parallel between learning and evolution. Behaviour 3: 174-215.

Pulliam, H.R. 1974. On the theory of optimal diets. Amer. Nat. 108: 59-79.

Raiffa, H. 1968. Decision Analysis. Addison-Wesley, Reading, M A.

Raven, C.P. 1961. Oogenesis: The Storage of Developmental Information. Pergamon Press, Oxford.

Reed, S.K. 1973. Psychological Processes in Pattern Recognition. Academic Press, N.Y.

Rhodes, J.L. 1971. Application of automata theory and algebra via the mathematical theory of complexity. Notes, Dept. Math., Univ. of California, Berkeley, California.

Robinson, J.E. 1975. Frequency analysis, sampling and errors in spatial data. (In) Display and Analysis of Spatial Data (J.C. Davis and M.J. McCullagh, eds.). J. Wiley and Sons, London.

Root, R.B. 1975. Some consequences of ecosystem texture. (In) Ecosystem Analysis and Prediction (S.A. Levin, ed.). Society for Industrial and Applied Mathematics, Philadelphia.

Rosen, B.K. 1974. Syntactic complexity. Information and Control 24: 305-335.

Rosenkrantz, D.J. 1967. Programmed grammars: a new device for generating formal languages. IEEE Ann. Symp. Switching and Automata Theory, 8th, Austin, Texas, Conf. Record.

Royama, T. 1970. Factors governing the hunting behavior and selection of food by the great tit (Parus major (L.)). J. Anim. Ecol. 39: 619-668.

Salzer, F. 1952. Structural Hearing: Tonal Coherence in Music. C. Boni, N.Y.

Schaeffer, W.M. 1974. Optimal reproductive effort in fluctuating environments. Amer. Nat. 108: 783-790.

Schleidt, W.M. 1973. Tonic communication: continual effects of discrete signs in animal communication systems. J. Theor. Biol. 42: 359-386.

Schoener, T.W. 1971. Theory of feeding strategies. Ann. Rev. Ecol. Syst. 2: 369-404.

Shaw, A.C. 1969. A formal picture description scheme as a basis for picture processing systems. Information and Control 14: 9-52.

Siegler, M.V.S., Mpitsos, G.J. and W.J. Davis. 1974. Motor organization and generation of rhythmic feeding output in buccal ganglion of Pleurobranchaea. J. Neurophysiology 37: 1173-1196.

Simon, J.C. 1975. Recent progress to formal approach of pattern recognition and scene analysis. Pattern Recog-

nition 7: 117-124.
Slater, P.J.B. 1973. Describing sequences of behavior. (In)
Perspectives in Ethology (P.P.G. Bateson and P.H. Klopfer,
eds.). Plenum Press, N.Y.
_____. 1974. The temporal pattern of feeding in the zebra
finch. Anim. Behav. 22: 506-515.
_____ and J.C. Ollason. 1972. The temporal pattern of behavior in isolated male zebra finches: transition analysis.
Behaviour 42: 248-269.
Slobodkin, L.B. 1968. Toward a predictive theory of evolution. (In) Population Biology and Evolution (R.C. Lewontin, ed.). Syracuse Univ. Press, Syracuse.
Smith, J.M.N. 1974a. The food searching behavior of two
European thrushes. I. Description and analysis of search
paths. Behaviour 48: 276-302.
_____. 1974b. The food searching behavior of two European
thrushes. II. The adaptiveness of the search patterns.
Behaviour 49: 1-61.
_____ and H.P.A. Sweatman. 1974. Food-searching behavior of
titmice in patchy environments. Ecology 55: 1216-1232.
Soule, S. 1974. Entropies of probabilistic grammars.
Information and Control 25: 57-74.
Strain, E.R. 1953. Establishment of an avoidance gradient
under latent-learning conditions. J. Exptl. Psych. 46:
391-399.
Tartara, G. 1973. On the application of algorithmic information theory to decision problems. Information Sciences
6: 85-96.
Taylor, H.M., Gourley, R.S., Lawrence, C.E. and R.S. Kaplan.
1974. Natural selection of life history attributes: an
analytical approach. Theoret. Popul. Biol. 5: 104-122.
Thatcher, J.W. 1970. Universality in the von Neumann cellular model. (In) Essays on Cellular Automata (A.W. Burks,
ed.). Univ. of Chicago Press, Chicago.
Thomason, M.G. 1973. Finite fuzzy automata, regular fuzzy
languages, and pattern recognition. Pattern Recognition
5: 383-390.
Thorpe, W.H. 1972. The comparison of vocal communication in
animals and man. (In) Non-Verbal Communication (R.A.
Hinde, ed.). Cambridge Univ. Press, Cambridge.
_____ and J. Hall-Craggs. 1976. Sound production and percep-
tion in birds as related to the general principles of
pattern perception. (In) Growing Points in Ethology
(P.P.G. Bateson and R.A. Hinde, eds.). Cambridge Univ.
Press, Cambridge.
Tobler, W.R. 1975. Linear operators applied to areal data.
(In) Display and Analysis of Spatial Data (J.C. Davis and
M.J. McCullagh, eds.). J. Wiley and Sons, London.
Tolman, E.C. 1948. Cognitive maps in rats and men. Psych.

Review 55: 189-203.

Treisman, M. 1975a. Predation and the evolution of gregariousness. I. Models for concealment and evasion. Anim. Behav. 23: 779-800.

_____. 1975b. Predation and the evolution of gregariousness. II. An economic model for prey-predator interaction. Anim. Behav. 23: 801-825.

Tsypkin, Y.Z. 1971. Adaptation and Learning in Automatic Systems. Academic Press, N.Y.

Uttal, W.R. (Ed.). 1972. Sensory Coding: Selected Readings. Little, Brown and Co., Boston.

van Hooff, J.A.R.A.M. 1970. A component analysis of the structure of the social behaviour of a semi-captive chimpanzee group. Experientia 26: 549-550.

von Neumann, J. 1966. Theory of Self-Reproducing Automata. Univ. of Illinois Press, Urbana.

Volkelt, H. 1912. Uber die Vorstellungen der Tiere: Ern Beitrage Z. Entwicklungspsychologie. Leipzig: Engelmann.

Vowles, D.M. 1970. Neuroethology, evolution and grammar. (In) Development and Evolution of Behavior (L.R. Aronson, E. Tobach, D.S. Lenrman and J.S. Rosenblatt, eds.). Freeman, San Francisco.

Wachtel, H. and E.R. Kandel. 1967. A direct synaptic connection mediating both excitation and inhibition. Science 158: 1206-1208.

Westman, R.S. In Review. An evolutionary bound to organismal complexity.

_____ and A.E. Dunham. In Review. A dynamical theory of resource utilization strategies.

Wharton, R.M. 1974. Approximate language identification. Information and Control 26: 236-255.

Wickler, W. 1968. Mimicry in Plants and Animals. McGraw-Hill, N.Y.

Wiepkema, P.R. 1961. An ethological analysis of the reproductive behavior of the bitterling. Arch. Neerl. Zool. 14: 103-199.

Will, B. 1974. Development of 'strategies' utilized by albino rats in operant conditioning. Anim. Behav. 22: 370-375.

Williams, K.L. 1975. A multidimensional approach to syntactic pattern recognition. Pattern Recognition 7: 125-137.

Wittgenstein, L. 1953. Philosophical Investigations. 2nd Ed. Macmillan, N.Y.

Wollheim, R. 1968. Art and its Objects. Harper and Row, N.Y.

Yost, J.L. 1973. Modelling a nerve cell membrane as a finite-state machine. M.A. thesis, Univ. California, Berkeley, C A.

Zadeh, L.A. 1965. Fuzzy sets. Information and Control 8: 338 ff.

_____. 1975a. The concept of a linguistic variable and its application to approximate reasoning. I. Information Sciences 8: 199-249.

_____. 1975b. The concept of a linguistic variable and its application to approximate reasoning. II. Information Sciences 8: 301-357.

_____. 1975c. The concept of a linguistic variable and its application to approximate reasoning. III. Information Sciences 9: 43-80.

Zeigler, B.P. 1976. Theory of Modelling and Simulation. J. Wiley and Sons, N.Y.

PARTITIONING DEPENDENCE
IN NONSTATIONARY BEHAVIORAL SEQUENCES

Neal Oden

University of Michigan

Abstract: A fault common to most methods of communication analysis is that they assume the data are stationary. A method is developed that does not assume stationarity and that allows assessment of the significance of dependence between acts separated by an arbitrary number of steps. The method essentially involves fitting ascending-order nonstationary Markov processes to the data. The statistic, based on a chi-squared distribution, is analogous to a partial correlation coefficient, and may be transformed to an information measure. Application of the method to records of aggressive communication in the hermit crab <u>Calcinus</u> <u>tibicen</u> reveals not only communication between individuals but the presence of an underlying motivational state.

INTRODUCTION

Many of the recent investigators of animal communication (Hazlett and Bossert, 1965; Dingle, 1969; Altmann, 1965) used transition matrices to describe their data. These matrices, constructed by enumeration of all observed one-step transitions, were subjected to chi-square analysis to reveal dependence between adjacent acts. Implicit in this method is the assumption that both the transition probabilities and the unconditional probabilities of single acts (the marginals of the transition matrix) are constant over the length of the interaction. If this assumption is met, the data are said to be stationary, and may be described by a Markov chain. A large body of theory has been developed around Markov chains (Feller, 1968; Kemeny and Snell, 1969) and the problems of fitting data to them (Goodman, 1962; Chatfield and Lemon, 1970; Coleman, 1964).

Unfortunately there is no reason to suppose that animal communication involves stationary probabilities. In aggressive interactions, for example, it seems reasonable that the ultimate loser and winner both change the probabilities with which they execute acts as the fight progresses. In this paper, a method is presented which analyses sequences of behaviors without assuming stationarity.

Hermit Crabs as an Experimental Subject

The ecology and behavior of the tropical hermit crab Calcinus tibicen has been investigated by Hazlett (1966). Adult hermit crabs normally live in empty gastropod shells, which are a limiting resource. Optimal shell sizes for each hermit crab size class are presumed to exist. Non-optimal shells may be either too small or too heavy (Fotheringham, 1976; Hazlett, 1970; Reese, 1963).

When individuals meet, they may either ignore each other (Hazlett, 1968) or, if similar in size, they may enter into the first phase of a 'shell fight' (Hazlett, 1966). This phase, which is marked by an exchange of visual signals, terminates when one of the crabs either leaves or ducks back into its shell. Occasionally, when a crab attempts to leave, the other crab holds on to the shell of the retreater, and then the retreater ducks back into its own shell.

If one crab does not retreat, phase 2 of the shell fight can begin. During this phase, the attacker repeatedly raps its shell on the shell of the defender, an action which may be repeated hundreds of times. In most species, no other physical contact occurs between the crabs at this time. The rapping finally is halted, either because the attacker leaves, or because the defender emerges from its shell, often climbing onto the shell of the attacker. A crab out of its shell is in a desperate situation. If it falls off the shell of the attacker and wanders away, its death is almost assured. In nature its life expectancy without a shell is on the order of ten minutes (Hazlett, person. comm.). After emergence of the defender, the attacker re-orients its body inside its shell, and while holding on to its shell, inserts its body into the shell of the defender. The fight is terminated when the attacker either occupies the new shell, releasing the old one and allowing the defender to enter it, or relinquishes the new shell, and re-enters its own.

During May 1975, some 40 well-matched individuals of Calcinus tibicen were subjected to an incomplete round-robin series of fights in the laboratory. In each case, the interactions were terminated by the experimenter at the end of Phase 1. This set of interactions, herein called a 'series', was replicated three times with different groups of crabs. Using a code discussed in Hazlett and Bossert (1965), the interactions were made machine-readable, and form the data base analysed in this paper.

These data are also being analysed by other methods, and a full description of experimental conditions will be given (Hazlett, in prep.).

I. TERMINOLOGY

"Act" and "step" are two words used repeatedly in this paper. In the description of any fight, act designates a particular behavior pattern or display performed, whereas step denotes the order in which acts occurred. Thus a fight may be described by saying that act C occurred in step 3, and so on. An "n-tuple" is a set of n adjacent acts or steps, depending on the context. A "series" is a set of round-robin matches.

II. INITIAL RESULTS OF ANALYSIS

To consider the question of stationarity one must examine the frequencies of various acts in the different steps. Table 1 shows the number of times that various acts were observed in steps 2-7 of 225 seven-step fights from series 3. Step 1 of all fights was act type A (approach) and this step is therefore omitted from the table. Since the number of steps in these fights is odd and the last act of a fight is a concession (retreat or duck in), all these fights were won

TABLE 1

The number of acts (A-Q) occurring in steps 2-7 of 225 seven-step fights in series 3. The first act is always A, and is not shown here. Even-numbered steps were performed by respondant-winners, and odd-numbered steps by initiator-losers.

STEPS	A	B	C	D	G	H	I	J	K	L	M	N	P	Q
2	173										52			
3	8	139		5				9	29		26	2	2	5
4	27	76							14	105			1	2
5	6	39			6			6	55		44	4	21	44
6	31	42	3		1				47	3	84		9	5
7						209	16							

by respondants. It seems unlikely that the probabilities of
acts or transition probabilities are stationary. Rather,
they appear to follow a pattern dependent on both time (beginning, middle, or end of fight), and the identity (initiator,
responder, winner, loser) of the actor. Table 2 shows similar
data for 158 six-step (initiator-won) fights in series 3.

In this context, it should be noted that anomalies exist
in the designations 'winner' and 'loser'. Since only phase I
of the fights was observed, the ultimate outcome of each interaction remains somewhat uncertain. An exchange between a
large crab with a small shell and a small crab with a large
shell is conceivable. In this situation, both animals
would be winners. If, on the other hand, phase 2 terminated
without the defender's emerging, the payoffs would also be
unclear. It should be noted, however, that the attacker is
less likely to lose than the defender since it retains the
option of reclaiming its own shell. For these reasons, this
paper takes 'loser' to be synonymous with 'the animal that
performs act H (retreat) or I (duck in)'. The 'winner' is
defined as the other animal.

Given that the unconditional probabilities of the acts
are not constant, it is still interesting to discover whether
the observed number of particular types of 1-step transitions
are significantly different from expected under the hypothesis
of independence. Consider that we are interested in a particular 2-tuple (= dyad) (pair of adjacent acts) which we de-
signate as A_1B_2. This designation should be taken to mean
act A occurred in step 1 and act B occurred in step 2. If we

TABLE 2

*The number of acts (A-Q) occurring in steps 2-6 of 158 six-
step fights in series 3. The first act is always A, and is
not shown here. Even-numbered steps were performed by respon-
dant-losers, and odd-numbered steps by initiator-winners.*

STEPS	A	B	C	D	G	H	I	J	K	L	M	N	P	Q
2	137								1		20			
3	9	82			2			4	38	16			3	4
4	17	35			4			3	14	80	1	1	3
5	9	23	1	1				3	67	20		23	11
6						103	55						

have a series of N fights of at least 2 steps, we can write our observation of the event A_1B_2 as a string of N digits, where the ith digit is 1 if the ith fight has the event, and 0 otherwise. If we take each of the digits as representing an independent outcome of an event with the probability P, the number of 1's in the string will be binomally distributed. If we have a null hypothesis that gives us P, we can test for the significance of the deviation from expected.

Under the null hypothesis of independence, $P(A_1B_2)$ should simply be equal to $P(A_1)P(B_2)$. Therefore, we can take as the estimate of $P(A_1B_2)$ under H_0:

$$\hat{P}(A_1B_2) = n(A_1)n(B_2)/N^2$$

where $n(A_1)$ means the number of times A occurred in the first step of the fight and N is the total number of fights.

This approach was carried out for every possible 2-tuple starting in every possible step. Note that all the data were not combined into one data set. Rather, a complete analysis was carried out for all fights of length 2, another for all fights of length 3 and so on. This restriction was adopted because it seemed reasonable that the probabilities would change at different rates in fights of different lengths (Hazlett and Estabrook, 1974a,b). Furthermore, if one is given enough data, it is always less dangerous to overcontrol than to undercontrol.

The statistical dangers inherent in this approach are twofold. In the first place, since the data are being assessed many times, we expect some large number of spuriously 'significant' findings. In the second place, there is co-variance between results in the sense that an unusually high number of one type of 2-tuple necessitates lower than expected numbers of other possible 2-tuples starting at the same step. For these reasons, the results discussed here should be taken as descriptions of the data rather than significant findings. An attempt to remedy these faults will be made in a later section of the paper.

Significant deviations from expected ($P \leq 0.05$, binomial test) were obtained from many subsets of the data. These results, which are not displayed here, were unclear, except for two things: (1) The 2-tuple M_iM_j almost always occurred significantly less than expected. (2) The act M figured in almost all of the significantly frequent and significantly infrequent 2-tuples.

In interpreting these findings we must take into account the fact that M stands for the act 'do nothing'. It is difficult to tell when one animal has stopped doing nothing and the other has started, and this doubtless explains the low M_iM_j frequency, but it does not explain the other finding.

Hazlett and Bossert (1965) found that 'do nothing' was a meaningful act in the agonistic interactions of this and other crustacean species.

A. Higher Order Interactions

An obvious extension of the method outlined above involves looking at 3-tuples, 4-tuples, and so on. To do this we might take as the probability of a specific type of 3-tuple:

$$P(A_1 B_2 C_3) = P(A_1)P(B_2)P(C_3).$$

However, if a particular 3-tuple turns out to be especially common, we won't know if this is the case because step 1 influences step 3, or because the 3-tuple happens to be composed of common 2-tuples. It is desirable to have a measure of how much step 1 affects step 3 after taking 1-2 and 2-3 transitions into account.

If dependence of adjacent steps has been demonstrated, it is clear that:

$$P(A_1 B_2) \neq P(A_1)P(B_2).$$

Since it is always true that:

$$P(A_1 B_2) = P(A_1)P(B_2 | A_1)$$

rejecting independence means that:

$$P(B_2 | A_1) \neq P(B_2).$$

In other words, A_1 has an effect on step 2. If we reject independence between adjacent acts, we have effectively asserted that at least 1-step transition probabilities are needed to adequately describe the data. To see if 1-step probabilities are enough, we note that, if no more than 1-step transition probabilities are needed,

$$P(A_1 B_2 C_3) = P(A_1)P(B_2 | A_1)P(C_3 | B_2).$$

It is always true that:

$$P(A_1 B_2 C_3) = P(A_1)P(B_2 | A_1)P(C_3 | A_1 B_2)$$

so in testing whether 1-step transitions are sufficient we are really testing whether:

$$P(C_3|B_2) = P(C_3|A_1B_2).$$

If not, A_1 has an effect on step 3 over and above the influence it exerts through step 2.

The method generalizes in an obvious way to higher-order dependencies. The assumption being tested, the prediction, and the meaning of rejection for some of the tests performed are outlined in Table 3. The actual estimators used to get the predictor probabilities follow in a natural way from the definitions of the predictor probabilities. For example, the estimator for $P(A_1B_2C_3)$ is:

$$\hat{P}(A_1B_2C_3) = \frac{n(A_1)}{N} \; \frac{n(A_1B_2)}{n(A_1)} \; \frac{n(B_2C_3)}{n(B_2)} = \frac{n(A_1B_2)n(B_2C_3)}{Nn(B_2)}.$$

For a given n-tuple and step-number combination, we can perform the same binomial test as before. The results for all length seven fights from series 3 are displayed in Table 4 with the expected and observed numbers of transitions, and the significance of the deviation under the binomial test. Only those transitions with $P \leq .1$ (binomial test) are listed. Again, 'M' figures importantly in the results.

The problems attendant upon looking at the data many times and upon the fact that there is covariance in the data were mentioned above. The fact that, for each n-tuple-step-

TABLE 3

Example of the Assumptions, Tests, and Meanings of Rejections for 2-tuples, 3-tuples, and 4-tuples.

Assumption	Prediction (Test)	Meaning of Rejection
Moves are independent	$P(A_1B_2)=P(A_1)P(B_2)$	Dependence between adjacent steps
At least 1-step transition probabilities needed	$P(A_1B_2C_3) =$ $P(A_1)P(B_2|A_1)P(C_3|B_2)$	Dependence from 2 steps away not transmitted through lower links
At least 2-step transition probabilities needed	$P(A_1B_2C_3D_4) =$ $P(A_1B_2)P(C_3|A_1B_2)$ $P(D_4|B_2C_3)$	Dependence from 3 steps away not transmitted through lower links

TABLE 4

Unusually frequent or infrequent n-tuples for 225 length 7 fights from series 3. Columns show starting step number, n-tuple, expected number assuming (n-2)-step transition probabilities suffice to describe data, observed number, one-tailed significance (binomial distribution), and whether deviation is positive or negative (minus sign means negative). Only P ≤ .1 deviations shown. Meaning of the Act code given in Appendix B.

Observed Transitions

STEP ACTS	EXP	OBS	SIG
1 MK	6.70	13	0.0183
1 MG	1.16	4	0.0297
1 AM	19.99	26	0.0984
1 AK	22.30	16	0.0979-
1 MJ	2.08	5	0.0593
1 MP	0.46	2	0.0788
1 MKA	3.59	8	0.0291
1 ABB	33.63	41	0.0993
1 MBM	13.67	20	0.0518
1 MBA	2.16	5	0.0671
1 MKMP	1.07	3	0.0935
1 MBMB	4.47	1	0.0607-
1 MBMQ	5.26	9	0.0842
2 KA	3.48	8	0.0249
2 BM	64.87	76	0.0588
2 KB	9.80	5	0.0708-
2 MB	8.78	21	0.0002
2 JA	1.08	3	0.0951
2 BBM	16.72	11	0.0921-
2 KMK	5.20	10	0.0378
2 ABJ	0.20	2	0.0170
2 JMF	0.67	3	0.0300
2 JMN	0.10	1	0.0909
2 AMJ	0.03	1	0.0282
2 NBN	0.03	1	0.0260
2 GBP	0.08	2	0.0029
2 NBN	8.57	17	0.0061
2 BBQ	8.09	14	0.0342
2 NBQ	4.14	1	0.0796-
2 BMP	10.13	3	0.0083-
2 NMM	0.02	1	0.0189
2 KKG	0.07	1	0.0689
2 QMP	0.40	2	0.0614
2 QMQB	0.10	1	0.0909
3 AB	4.68	8	0.0999
3 BM	14.86	31	0.0000
3 KM	2.74	7	0.0212
3 MK	25.67	39	0.0036
3 BK	18.58	9	0.0139-
3 KB	2.43	5	0.0982
3 BB	13.17	7	0.0536-
3 BP	7.09	3	0.0738-
3 NM	20.53	2	0.0000-
3 BKB	0.82	3	0.0498
3 QKQ	0.04	1	0.0357
3 KBK	1.03	4	0.0203
3 BQK	2.39	5	0.0931
3 BQM	6.82	3	0.0884-
3 MQM	9.55	15	0.0578
3 PBP	0.08	1	0.0741
4 MK	9.19	21	0.0004
4 KM	20.53	30	0.0190
4 MP	1.76	6	0.0091
4 KB	10.27	5	0.0537-
4 KK	11.49	6	0.0654-
4 MQ	0.98	3	0.0759
4 PB	3.92	1	0.0956-
4 GAI	0.06	1	0.0625
5 KI	3.34	9	0.0070
5 MI	5.97	1	0.0168-
5 CI	0.21	2	0.0197

Unobserved Transitions

STEP ACTS	EXP	OBS	SIG
1 MM	6.01	0	0.0023-
2 MM	12.13	0	0.0003-
3 KK	3.42	0	0.0318-
4 MM	16.43	0	0.0000-
5 BI	2.99	0	0.0495-
1 MBB	7.37	0	0.0006-
1 AKA	4.41	0	0.0116-
3 MKB	3.55	0	0.0281-

number in a series, a complete list of observed and expected frequencies is available permits a chi-squared test that allows us to estimate the significance of the deviation from expected for that n-tuple-step-number, and avoids some of the statistical difficulties. However, the correct number of degrees of freedom for the test is not obvious.

Suppose, for example, we are investigating all the 3-tuples that can arise starting in step 1 of a set of interactions where only two acts, A and B, are possible in each step. One way to arrange the observed counts is depicted in Figure 1. Here, the row names consist of the set of all possible

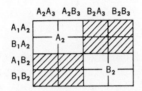

Fig. 1. A table suitable for recording all 3-tuples starting in step 1 of a set of interactions, where A and B are the only acts possible in each step. Boxes containing structural zeros are shaded. The table is separable into subtables, each containing all the data for a unique step 2 move.

2-tuples starting in step 1, and the column names consist of the set of all possible 2-tuples ending in step 3. Certain boxes, shaded, in the figure, correspond to impossible situations. For example, it is impossible to have A_1A_2 followed by B_2A_3. The unshaded boxes form a set of nonoverlapping subtables, each subtable depicting all and only the transitions that go through a specified step 2 act.

In Figure 2, the subtable dealing with all acts containing B_2 has been excised, and displayed with its marginal counts and grand total. Suppose we are trying to estimate the expected frequency for the 3-tuple defined by the box containing the X. From before, the expected frequency is:

$$P(A_1B_2A_3) = P(A_1)P(B_2|A_1)P(A_3|B_2)$$

and its estimator is:

$$\hat{P}(A_1B_2A_3) = \frac{n(A_1B_2)n(B_2A_3)}{Nn(B_2)}$$

Therefore the expected frequency of $A_1B_2A_3$ is:

X		$n(A_1B_2)$
		$n(B_1B_2)$
$n(B_2A_3)$	$n(B_2B_3)$	$n(B_2)$

Fig. 2. The lower right subtable of Fig 1, showing marginal counts and grand total.

$$\frac{n(A_1B_2)n(B_2A_3)}{n(B_2)}$$

A glance at Figure 2 will reveal that this estimator is exactly the one that would be used if performing an ordinary chi-squared test of independence between the rows and columns of Figure 2. This is true of all the other estimators.

On the strength of this argument one can conclude that the overall chi-squared calculation is equivalent to the sum of the chi-squared calculations for each of the subtables, and the appropriate degrees of freedom is simply the sum of the degrees of freedom for the separate subtables. Each subtable holds constant a unique (n-2)-tuple and has as row names the names of starting acts, and as column names the names of ending acts. This indicates that for every n-tuple, we are testing whether the probability of the last act, given the preceding n-2-tuple, is the same as the probability of the last act given the preceding n-1-tuple.

Although it is customary to use Pearson's X^2 approximation (the sum of the squared deviations divided by the expecteds) to perform the chi-squared test, I have elected to use not this, but rather, two other tests.

The first of these is the G-test, or likelihood-ratio test:

$$G = 2\Sigma o_i \ln \frac{o_i}{e_i}$$

where o_i and e_i are the observed and expected frequencies in the ith cell. For an n x m table, G is distributed as a chi-square with n-1 x m-1 degrees of freedom.

The second test involves Haldane's calculation for the mean and variance of Pearson's X^2 (Haldane, 1939). A widespread problem in the study of animal communication is that, typically, although an investigator has many observations, the table in which these observations reside is very large, with the result that many cells have low expected counts. In this case, the value of Pearson's X^2, the G-statistic, and also the amount of information in common between the rows and columns,

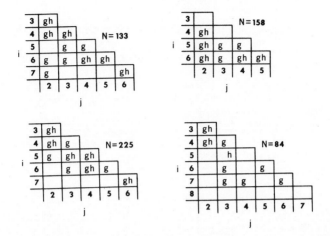

*Fig. 3. A g or h in cell ij indicates that the approp-
riate test (the G-test or Haldane's modification of X²) shows
that step i is dependant on step j over and above whatever
influence is passed through lower linkages.*

upper left   - series 2 length 7
upper right - series 3 length 6
lower left   - series 3 length 7
lower right - series 3 length 8

are all seriously overinflated (see Steinberg, this volume).
Haldane has calculated exactly the first two moments of
Pearson's X^2, and if the sample size is large, X^2 can be
treated as a normal deviate, with Haldane's mean and variance
(Cochran, 1954). This is true independent of the expected
cell counts. The Central Limit Theorem assures us that, even
if the sample size for each subtable is small, the sum of the
subtable X^2 values will still approach a normal distribution.
The mean and variance will equal the sums of the means and
variances of the subtables.

    The results of these statistical tests are presented for
a number of different fight lengths in Figure 3. In these
diagrams, the presence of a g or h in cell ij indicates that
the G-test or Haldane's modification of X^2 rejected independence between steps j and i; step i was dependent on step j
over and above whatever influence was transmitted through
intervening steps.

    As expected, in most cases, Haldane's is more conserva-
tive than the G-test. The results of the tests suggest that
usually no more than two-step transition probabilities need be
invoked to adequately describe the data. This conclusion is
equivalent to the assertion that no more than a second order

nonstationary Markov process need be postulated. It appears
that a hermit crab not only responds to the act of its
opponent, but also that its acts reflect an underlying motivational state, i.e., what a crab does is not independent of
what it did previously.

   A chi-squared test can be thought of as measuring the
significance of the distance between the expected cell distributions and the observed cell distributions. The significance depends not only on the distance but on the number of
observations. It is sometimes interesting to obtain a measure
of the simple distance between these two populations. If the
null hypothesis is that rows and columns are independent, the
distance between expected and observed cell counts also
serves as a measure of association between the rows and
columns. The X^2 or G value divided by the sample size provides a measure of association. The Shannon-Weaver expression
for the amount of information held in common between the rows
and columns of a matrix is:

$$I = \Sigma \; P_{ij} \; \log_2 \; (P_{ij}/(P_{i+}P_{+j}))$$

where P_{ij} stands for the probability of an observation in cell
ij, P_{i+} stands for the sum of the probabilities in the ith
row, and so on. If one replaces the probabilities by their
estimators, one obtains:

$$\hat{I} = \frac{1}{N} \Sigma \; o_{ij} \; \log_2 \frac{No_{ij}}{o_{i+}o_{+j}}$$

where o_{ij} stands for the number of observations in cell ij
and $\hat{I}$ is an estimate of I.

   The G-statistic for independence between rows and columns
of a matrix is:

$$G = 2 \; \Sigma \; o_{ij} \; \ln \frac{No_{ij}}{o_{i+}o_{+j}}$$

   It is clear that, under the hypothesis of independence,
$K\hat{I}$ is distributed as a chi-squared variable with $(r-1)(c-1)$
degrees of freedom, where $K = 2N\ln2$. Since it is distributed
as a chi-square variable, if N is large, the expected value of
I is given by

$$E(I) = \frac{\nu}{K}$$

where ν is the degrees of freedom, and the variance of I is
given by

$$V(I) = \frac{2\nu}{(K)^2}$$

Furthermore, it is possible, by dividing each of the G-tests discussed above by K to turn it into an information-theoretic measure of association.

An argument is present in the appendix that clarifies the meaning of these measures. Briefly, if one symbolizes the uncertainty in a collection of events W as H(W) and the uncertainty in another collection X as H(X), one can consider the amount of information in common between the two sets I(W;X) to be the difference between H(X) and the uncertainty in X, given that one knows which event in W has happened. In symbols:

$$I(W;X) = H(X) - H(X,W).$$

This relation may be represented by the diagram shown in Figure 4, where one circle represents the uncertainty in W, the other the uncertainty in X, and the overlap represents the information in common. If the two sets are independent then I(W;X) = 0. Let W represent all possible step 1 acts and X represent all possible step 2 acts. Then I(W;X) is that quantity obtained by dividing the statistic for 2-tuples by K. For three sets of events, a similar diagram may be drawn. Let Y represent the set of all possible step 3 acts. For each act occurring in step 2, we may form a table to test independence of step 1 from step 3, as indicated above and shown in Figure 4. We can sum the G-statistic for all such tables and this quantity is here referred to as the G-statistic for

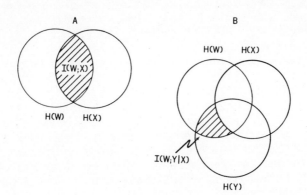

Fig. 4.   Venn diagram illustrations of information-theoretic quantities.

A - shaded portion represents amount of information shared by sets W and X.

B - shaded portion represents amount of information shared by sets W and Y that is not also shared by X.

3-tuples. If it is divided by K it is the amount of information in common between W and Y that is not shared with X. This quantity, which is not $I(W;Y)$ but rather $I(W;Y|X)$ is shaded in Figure 4B. Similar relations hold for higher n-tuples; in each case, the G-statistic is transformed into the amount of information shared between the first and the last steps of the n-tuple, but not with any other steps (see Appendix A). This makes it clear that the G-statistics calculated here are similar to partial correlation coefficients.

This approach is similar to that of Chatfield and Lemon (1970). The account of information theory presented here is woefully inadequate and the reader is urged to consult Abramson's (1963) introduction.

Maynard Smith and Parker (1976) introduced the concept of a "war of attrition". In this type of contest, each animal has a secret time limit, beyond which it refuses to fight. All fights are therefore won by the animal with the larger limit. These authors showed that this strategy can be evolutionarily stable in the sense that, for certain distributions of time limits, mutants that don't "play by the rules" will be selected against.

An important point about this sort of contest is that it entails no communication between the actors. Rather, they are completely pre-programmed to obey an internal clock, and, if there is systematic variation in their actions, it is entirely due to the structure of the clock.[1] If one were to assume stationarity in inspecting the actions of these animals, one would probably find spurious deviations of n-tuple frequencies from expected, and would falsely conclude that the animals were communicating. The nonstationarity method described above assumes the clock is operating. Therefore if this method discovers communication (as it has for these hermit crab data), the demonstration is stronger than if the communication is demonstrated using stationary methods.

Whether or not the "clock" model actually occurs in nature, one may at least conceive it is the endpoint of a continuum ranging from no communication to 'complete communication', a situation where each act depends completely and only upon the act of the opponent. The results of this paper suggest that hermit crab fights have both a 'clock' and a 'complete communication' component.

---

1. Editors footnote: While a model predicting non-communication during fights between animals may seem unlikely to ethologists looking for communication, such a situation has been reported in the grasshopper Dissosteira carolina (Kerr, person. comm.).

ACKNOWLEDGMENTS

The creative suggestions and critical comments of Drs. Brian Hazlett, George Estabrook, and William Du Mouchel are gratefully acknowledged. Discussions with Dr. Peter Oden on the nature of 3-way dependence proved very illuminating. The graphs and tabular material prepared by Jane Oden aided the analysis substantially. The Division of Biological Sciences at the University of Michigan generously donated computer time.

The responsibility for any errors included in this manuscript remains entirely my own.

APPENDIX A

Showing the relation between the G-tests of predictor probabilities and various information measures.

Given the set of adjacent steps W,X,Y,Z... define the family of information measures:

$$T_1(WX) = H(X) - H(X|W) = \Sigma\Sigma\ P_{wx}\ \log_2 \frac{P_{wx}}{P_w\ P_x}$$

$$T_2(WY) = H(Y|X) - H(Y|WX) = \Sigma\Sigma\Sigma\ P_{wxy}\ \log_2 \frac{P_{wxy}\ P_x}{P_{wx}\ P_{xy}}$$

$$T_3(WZ) = H(Z|XY) - H(Z|WXY) = \Sigma\Sigma\Sigma\Sigma\ P_{wxyz}\ \log_2 \frac{P_{wxyz}\ P_{xy}}{P_{wxy}\ P_{xyz}}$$

etc.
It can easily be verified that T_1 and T_2 are the quantities shaded in the Venn diagrams of Figure 4. These quantities are:

$$\hat{T}_1(WX) = \frac{1}{N}\ \Sigma\Sigma\ o_{wx}\ \log_2 \frac{No_{wx}}{o_w o_x}$$

$$\hat{T}_2(WY) = \frac{1}{N}\ \Sigma\Sigma\Sigma\ o_{wxy}\ \log_2 \frac{o_{xyz}\ o_y}{o_{xy}\ o_{yz}}$$

$$\hat{T}_3(WZ) = \frac{1}{N}\ \Sigma\Sigma\Sigma\Sigma\ o_{wxyz}\ \log_2 \frac{o_{wxyz}\ o_{xy}}{o_{wxy}\ o_{xyz}}$$

etc.
Now, in testing the fits of increasing orders of Markov chains, the predictor probabilities were:

$$P(wx) = P(w)P(x)$$

$$P(wxy) = \frac{P(wx)P(xy)}{P(x)}$$

$$P(wxyz) = \frac{P(wxy)P(xyz)}{P(xy)}$$

etc.

That is, the expected counts for the various types of tuples are:

$$e_{wx} = \frac{o_w\,o_x}{N}$$

$$e_{wxy} = \frac{o_{wx}\,o_{xy}}{o_x}$$

$$e_{wxyz} = \frac{o_{wxy}\,o_{xyz}}{o_{xy}}$$

etc.

The G-statistic calculated for the various tuples are:

$$G_1 = 2\,\Sigma\Sigma\,o_{wx}\,\ln\frac{No_{wx}}{o_w o_x}$$

$$G_2 = 2\,\Sigma\Sigma\Sigma\,o_{wxy}\,\ln\frac{o_{wxy}\,o_x}{o_{wx}\,o_{xy}}$$

$$G_3 = 2\,\Sigma\Sigma\Sigma\,o_{wxyz}\,\ln\frac{o_{wxyz}\,o_{xy}}{o_{wxy}\,o_{xyz}}$$

etc.

From this it is clear that:

$$2N\ln 2\,T_i = G_i.$$

APPENDIX B

Meaning of the act code utilized in this study. Descriptions of the behavior patterns are given in Hazlett (1966).

Letter	Behavior Pattern
A	approach
B	double cheliped presentation
C	double cheliped extension
D	single ambulatory raise
G	high body posture

H	retreat
I	duck into shell
J	high body posture while approaching
K	double cheliped presentation while approaching
L	double cheliped extension while approaching
M	do nothing, apparently "ignore"
N	displacement feeding
P	high body posture and double cheliped presentation while approaching
Q	high body posture and double cheliped presentation.

REFERENCES

Abramson, N. 1963. Information Theory and Coding. McGraw-Hill, N.Y.

Altmann, S.A. 1965. Sociobiology of rhesus monkeys. II: Stochastics of social communication. J. Theor. Biol. 8: 490-522.

Chatfield, C. and R.E. Lemon. 1970. Analysing sequences of behavioral events. J. Theor. Biol. 29: 427-445.

Cochran, W.G. 1954. Some methods for strengthening the common Chi-square tests. Biometrics 10: 417-451.

Coleman, J.S. 1964. Introduction of Mathematical Sociology. Collier-Macmillan Canada Ltd.

Dingle, H.A. 1969. Statistical and information analysis of aggressive communication in the Mantis Shrimp Gonodactylus bredini Manning. Anim. Behav. 17: 561-575.

Feller, W. 1968. An Introduction to Probability Theory and Its Applications, Vol. 1. J. Wiley & Sons, Inc. New York.

Fotheringham, N. 1976. Population consequences of shell utilization by hermit crabs. Ecology 57: 570-578.

Goodman, L.A. 1962. Statistical methods for analysing processes of change. Amer. J. Soc. 68: 57-78.

Haldane, J.B.S. 1939. The mean and variance of X^2 when used as a test of homogeneity, when expectations are small. Biometrika 31: 346-355.

Hazlett, B.A. 1966. Social behavior of the Paguridae and Diogenidae of Curacao. Stud. Fauna Curacao 23: 1-143.

_____. 1968. Size relationships and aggressive behavior in the hermit crab Clibanarius vittatus. Z. Tierpsychol. 25: 608-614.

_____. 1970. Tactile stimuli in the social behavior of Pagarus bernhardus (Decapoda, Paguridae). Behaviour 36: 20-48.

_____ and W.H. Bossert. 1965. A statistical analysis of the aggressive communications systems of some hermit crabs. Anim. Behav. 13: 357-373.

220   NEAL ODEN

_____ and G. Estabrook. 1974a. Examination of agonistic
behavior by character analysis. I. The spider crab
Microphrys *bicornutus*. Behavior 48: 131-144.
_____ and _____. 1974b. Examination of agonistic behavior by
character analysis. II. Hermit crabs. Behavior 49:
88-110.
Kemeny, J.G. and J.L. Snell. 1969. *Finite* *Markov* *Chains*.
D. van Nostrand Co., Inc., Amsterdam.
Maynard Smith, J. and G.A. Parker. 1976. The logic of
asymmetric contests. Anim. Behav. 24: 159-175.
Reese, E.S. 1963. The behavioral mechanisms underlying shell
selection by hermit crabs. Behaviour 21: 78-126.

Index